Smart Citizens in Smart Cities

The Fourth Industrial Revolution (Industry 4.0)

Smart Citizens in Smart Cities

The Fourth Industrial Revolution (Industry 4.0)

by Malik Luti

Library of Congress Control Number:
TXu002262689 / 2021-05-29

ISBN:	Paperback	978-1-7375117-4-8
	Ebook	978-1-7375117-1-7
	Audio	978-1-7375117-2-4

This book was printed in the United States of America.

Dedication

To the memory of my mother, who never had the chance to learn how to read or write. She is there somewhere in Heaven, smiling and happy to see me publishing this book.

Contents

Acknowledgments

This book is the result of several years of research. I want to express my gratitude for all the help and support I received from everyone around me, with special thanks to Bob Duncan, Paul Weisser, Bilal Elkaissi and four exceptional undergraduate students: Alex Ramirez, Aaron Ramirez, Luis Rodriguez, and Lucas Colias.

About the Author

Malik Luti is an enterprise strategic architect at The Boeing Company. He also teaches computer science at California State University, Long Beach. In addition, he lectures on enterprise architectural practices for governmental and non-governmental agencies within the United States and Canada. His clients have included the U.S. Department of Defense (DOD), the Canadian Department of National Defense (DND), the Royal Canadian Mounted Police–BC, Shared Services Canada (SSC), Public Services and Procurement Canada (PSPC), AT&T, Lockheed Martin, Intel, Verizon, VISA, Sony, J.P. Morgan, Bank of America, and many more.

Malik also provides guidance, motivation, and emotional support by coaching fellow workers and guiding newly hired employees and graduate students in finding their career paths. He considers his role as a mentor a privilege as well as deeply rewarding. It is in that same spirit that he was moved to publish this book.

Introduction

The size of the classes that I teach at Cal State Long Beach usually number up to thirty-five students. In some cases, the number reaches thirty-seven or thirty-eight. For the advanced levels, I usually don't extend the attendance. However, for the beginner classes, I sometimes do.

One time I was assigned to teach an introductory class that had almost 200 students. On the first day, while walking to my class, the thought occurred to me, *What's the best way to take attendance for a class that's filling up one of the largest lecture halls on campus, and how long will it take?*

Normally, with thirty-five students, it takes about eight to ten minutes to run down a list of names and notate any absences. Taking roll call for a class of that size is manageable, but for an attendance of 200 or more, it can take up to forty-five minutes. Forty-five minutes from a one-hour lecture: how is that efficient?

I thought of circulating a sheet of paper for the students to list their names on while class was in session. But then I realized that some students might list an absent friend's name for credit, since I had made it clear that missing 25 percent of classes would result in a failed grade.

Like most schools in the country, California State has an online system for both students and instructors. The system not only manages grades, but also uploads course contents, class schedules, payroll, and so on. However, it is not very useful for taking attendance, since that has to be done manually for 200 students. You can imagine how long that can take. Although the school does assign at least one teaching assistant to each of the professors, that still does not help the time issue, because canvassing the room for attendance during a lecture can be highly disruptive.

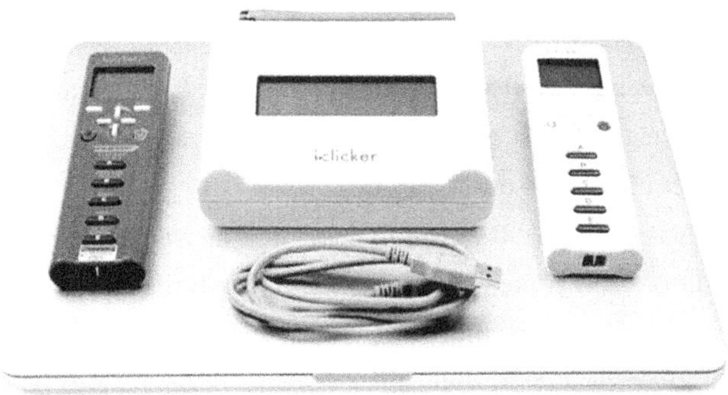

At one point, the school partnered with a company called i-Clicker to help with this issue. I then only needed to submit a request, and I was furnished with an i-Clicker server. The students were given links to purchase the i-Clicker remote devices, which were priced at $27.99. The devices were simple, with only five lettered buttons—*A*, *B*, *C*, *D*, and *E*, as shown in the illustration.

After configuring the remotes, the students were connected to my i-Clicker server, which was limited to my classroom. Once they were connected, I could ask them to click *A* if they were present. From then on, with only one click from each student, accurate class attendance was accounted for. As a bonus, I was also able to use the device to put true or false questions in my slides, and could ask the students to click *B* for a true answer and *C* for a false one.

This new timesaving device was a classic example of how digitization made life easier for both my students and me.

One of the most distinctive traits that set human beings apart from other species is that we are creative at problem-solving. Throughout history, the industrial revolutions bear witness to our inherent ability to solve the problems of speeding up production. Each new wave of industrial change profoundly affected the course of our lives.

The First Industrial Revolution not only ushered in technological advancements, but also brought about new social changes, as well as demographic and economic advancements. We are currently living in yet another industrial revolution, known as "Industry 4.0," which is having nothing less than a seismic impact on our lives.

Because of that revolution's all-encompassing impact, learning about it will help us not only to better understand the past, but also to come to grips with the world that we are living in today. The more knowledge we have of Industry 4.0's inner workings, the clearer our forecasts of the future will be.

This historic transformation involves all of us, and it is for that reason that I believe this information should be accessible to as wide an audience as possible. Everyone from high school students to employees, managers, and executives should be made aware of what is going on.

This book is divided into three parts:

Part 1 is a review of the various industrial revolutions that have taken place and how they evolved over time.

Part 2 includes examples of policies and policymakers today and how markedly unaligned they are with the current technological advances of this Fourth Industrial Revolution,

Industry 4.0. I have highlighted some points that reveal the subsequent impact that this revolution is making on all of our lives.

Part 3, which emphasizes the importance of actively engaging in our communities, illustrates how our more traditional role as passive residents of our communities is no longer tenable.

Part 1
The New Industrial Revolution

We are currently living in the Fourth Industrial Revolution, which is affecting our lives in unanticipated ways. Technology has become embedded within our societies, but feels like an exogenous force with the power of a tsunami. However, all of these emerging technologies are only a reflection of our desires and choices.

In part, I am going to explain why many companies and organizations have recently undergone complete transformations in order to stay afloat. I will briefly review the history of the industrial revolutions that preceded this one and how each one led to the next. Most importantly, I want to enlighten you, the reader.

Francis Bacon said it best: "Knowledge is power."

Speaking in the current vernacular, consider this information a much overdue update.

Chapter 1
Technology Is Changing Business

Digital Transformation

Digital transformation refers to the adoption of digital technology to transform services or business processes, thereby achieving a more efficient, effective, and competent means of meeting market demands. The digital transformation now taking place is so all-encompassing that it is fundamentally changing how we do business.

Digital transformation aims to integrate data from multiple data sources to automate as many business processes as possible. Of course, that requires utilizing technologies, such as robotics, artificial intelligence, machine learning algorithms, and so on.

Leadership and culture play a pivotal role in the success of this transformation. Throughout this book, I have provided several examples of governmental and non-governmental leadership in the United States and other nations that are in the midst of digitally transforming their infrastructures. Some of these attempts have been successful, and others have not. For the failures, it is important for us to be aware of which of these can be attributed to inept leadership.

Digital transformation is not only correlated with

change, but also with disruption. It has revitalized, dismissed, rejuvenated, discarded, revamped, created, and in many cases eliminated long-standing components of the business world. It has set forth new rules and regulations about how businesses now operate. It has irked both employers and employees.

Since time immemorial, change itself has been met with a mixture of doubt, fear, and anxiety. Quips like "We don't do it that way," "That's not our real focus," or "I don't think that's going to work" are some of the objections that are regularly heard against change. By reframing a given situation into a correct context, however, competent leaders can quell such comments. It is incumbent on everyone to ensure that all of our leaders are technologically savvy, so that they may more ably navigate the turbulent changes ahead.

Referring back to the i-Clicker: although it was considered cutting-edge technology ten or fifteen years ago, by today's standards it is inefficient. Consider the time used by both the professor and the teaching assistant in operating it. For example, the steps required to process the i-Clicker included: connecting the device's server to a laptop; downloading the data to a computer; and then manually entering that data line by line into the school's system. Also, the remote devices needed regular battery replacements, and at the end of the term the remotes became obsolete, unless the students were able to resell them. Using an i-Clicker was preferable to manual procedures, but a hardware device like that has now been replaced by an app that can do the same job faster and cheaper.

Apps not only provide a better user experience, but they also cut costs, run faster, and in some cases maintain a cleaner environment by reducing the use of plastics and unneeded hardware. Any learning institution can now easily integrate apps within its content management systems, and students can easily download those apps from an AppStore. Because of that integration and automation, the data can now be automatically uploaded from students' phones to the school's systems without any human intervention. Additionally, as an instructor, I can see the results in real time, which not only cut down on the use of paper and unnecessary hardware, but also save students from paying the fees required to purchase or maintain the devices.

Given the prevalence of digital transformation, the manufacturer of the i-Clicker completely altered its business model by developing an app, which it is now promoting. One alternative for schools would be to partner with the manufacturer of the i-Clicker to lease the app as a service for monthly or annual fees. Another alternative would be for the school to develop an app that can perform the same function, which would eliminate the need for a contract with an outside resource.

After undergoing a digital transformation, i-Clicker became a more cost-effective and efficient service for end-users, but the company itself had to pare down its staff, since many factory and team workers were no longer needed. What had formerly been a hardware-based end-product underwent a metamorphosis into an app. Assembly line workers were replaced by app developers, a job description that didn't even exist a few years ago! That change was crucial for business survival; otherwise, the company would have fallen behind, and its technology would have been rendered obsolete.

The i-Clicker company is only one of countless businesses that have had to entirely reinvent themselves in the past few years. As end-users continually look to improve their performance output and to cut costs, businesses the world over continue morphing into previously unimaginable dimensions. According to the International Data Corporation, digital transformation will grow at a compound annual growth rate of 15.5 percent between 2020 and 2023, and is expected to approach $6.8 trillion as companies build on existing strategies and investments to become digital enterprises (Murray, 2020).

The digital transformation is upon us. We have become like the passengers in H. G. Wells's *Men Like Gods*, who find themselves transported into another world—a world of ever-increasing speed and advances.

We are all in this together, and availing ourselves of these new technological and digital wonders will better serve us if we don't resist them. Historically, resisting the new has proven time and again to be futile.

Everyone in the Race

In July 2019, the U.S. Air Force published a timely article on its website entitled "The Digital Air Force" (U.S. Air Force, 2019), which stated that our world is entering a new age of technological discovery and advancement, owing to the use of data analytics and sensors. The article went on to delineate exactly how these advances are changing our communities, businesses, and military forces.

Air Force leaders have come to realize that we are in the midst of a technological revolution. Achieving victory in combat, for instance, can no longer depend on the capabilities of individuals. Robots and interconnected networks of sensors and analytic tools are performing tasks formerly done by humans. These artificial intelligence methods now empower airmen to make decisions faster and with greater efficiency.

The Air Force's new strategy has streamlined and integrated informational systems, data management, and business operations into one cohesive network. The initiative overhauled the information technology infrastructure that is responsive to the demands of combat, while simultaneously integrating data from a variety of sources, which in turn automated differentiated manual tasks. The new strategy has also meant that the very process of recruitment has been improved, as well as the training of Air Force staff, both of which necessitated a cultural change within the Air Force.

The Ford Motor Corporation is another example of an organization going through a digital transformation. In an interview in 2017, William C. Ford, Jr., the chairman of Ford, was quoted in the *New York Times* as saying, "Our world has changed dramatically. Look at the pace of change and the competitors coming into our space. We need to match or beat that" (Boudette, 2020). He then introduced Jim Hackett as the new CEO of Ford, who announced the company's digital transformation strategy within the first 100 days of his appointment. Hackett stated that Ford would embrace multiple technologies and focus on "integrating hardware and software in complex devices." An announcement of this nature was uncharacteristic of the Ford Motor Company. In one fell swoop, the plan mobilized Ford forward from its outmoded manufacturing approach to a digital approach in which its "smart" factories are powered by robotics, machine learning, and 3D printing. The stratagem not only cut costs, but also resulted in a smaller manufacturing footprint. The company projected that the new policy would cut:

vehicle development time by 20%
automotive growth cost through 2022 by 50%
reduction in footprint by 50%
reduction in capital investment by 50% (Dignan, 2017)

To accelerate this transformation, the Ford Motor Company signed a six-year partnership with Google in 2021, which is part of an $11 billion investment by Ford to upgrade its infrastructure.

Almost all businesses and organizational entities, whether governmental or civilian, have now undergone or are currently undergoing some form of digital transformation. That is an observable fact that is occurring in all areas of our lives. It can be seen in a myriad of shapes and forms, such as smart homes, smart cities, smart factories, smart buildings, and so on.

Is it possible, however, that this digital transformation is part of something even bigger than what meets the eye? How are our day-to-day lives being impacted by these changes? Glancing back to see how technology has affected our lives since the Renaissance reveals a great deal. In the next chapter, I will look at the history of the advancing industrial revolutions and discuss how they started and where they are now.

Chapter 2
Four Industrial Revolutions

The Birth of the Industrial Revolution

Technological revolutions have been occurring throughout the centuries, replacing one methodology with another, and in increasingly shorter periods of time. Although most advances were initially met with resistance, they undeniably brought about significant improvements, reshaping society and ultimately entire cultures by stepping up productivity and efficiency.

The Renaissance brought about two major social changes: capitalism and religious reform (History, 2020). The first change came about with the emergence of private ownership as a means for generating profits within free and competitive markets. That change helped to set the stage for the beginning of capitalism. The second change came along with the Reformation, during which, for the first time, parishioners began to individuate from the religious dogma that had long held sway over their lives.

It was during the reawakening of the Reformation that thinking minds began to look to science and technology rather than religion for answers to basic questions about life. The Catholic Church denounced such independent thought as heretical. However, by the seventeenth century, science had taken hold and was here to stay. The Age of Discovery was afoot, setting in motion a "scientific revolution" that has been with us ever since.

In the sixteenth century, the Dutch emerged triumphant over the Spanish and Portuguese with the dawning of the Dutch Golden Age. Much of what fueled the Netherlands' stratospheric rise in profits was due to that nation's innovative means of production and free trade practices. In a relatively short time, the Netherlands became the epicenter of commerce. With shipbuilding yards and sugar refineries thriving from limitless reserves of low-cost windmill energy, and with improved canal transport, the country's economy boomed.

The ever-increasing gravitational force of the Netherlands continued to attract more and more skilled workers into its orbit. Yet, despite increased production and continuous output of wood-crafted products, the Dutch soon found themselves unable to satisfy the staggering demands of the very market they had helped to forge. The great forward thrust that had come from energy sources like water, wind, and peat had begun to lose momentum.

Since the middle of the sixteenth century, British farmers have incrementally increased their annual output. Although this gradual buildup has been termed the British Agricultural Revolution, the progress responsible was entirely derived from the technological pioneering of the Dutch.

Britain's coal, however, fueled massive technological advances, making it the preeminent source of energy. By the close of the seventeenth century, the British Empire had taken the lead in manufacturing goods produced by machinery that ran on coal. In time, paying the highly skilled workers that were needed to run the factories increased wages. As a result, the income of the average British household increased, and with that increase, a new market for non-essential commodities emerged. For the first time, there was a demand for gourmet food, fashionable clothing, oil paintings, and a plethora of delicacies and fineries that never before had been within the reach of so many. The upward rise of the new market soon reached dizzying heights, which brought forth an unprecedented number of employment opportunities to keep the economy going.

Yet, even with a huge workforce in place, the escalating orders for quantity and quality called for ever more innovative methods to keep pace. Enter Thomas Newcomen and the first steam engine. In 1712, Newcomen introduced a remarkable invention, which, lo and behold, harnessed the energy of steam. That remarkable contraption spearheaded the Industrial Revolution.

For a long time, owing to low labor costs, Indian textiles controlled the lion's share of the world's cotton market. That arrangement consistently weakened Britain's place in the textile market, since British fabrics primarily consisted of costlier wools and linens. Unable to compete, the British government threatened India's economy by imposing a wholesale ban on all imported cotton textiles from India. Over time, the unyielding restrictions created an insatiable demand in Britain for all things cotton.

This was the environment that inspired John Kay, a clockmaker, to introduce a revolutionary device in 1734 that greatly expedited the production of cotton broadcloth. Known as the "flying shuttle," it enabled a single laborer to operate a loom. Allowing the shuttle carrying the weft to be passed through the warp threads faster and over a greater width of cloth greatly accelerated the process of weaving (History,

2020). Unfortunately, Kay's shuttle had some limitations and therefore was not widely adopted. However, it knocked over the first domino in a succession of innovations, which led to the fully industrialized output of products that we see today.

In 1769, Richard Arkwright partnered with John Kay to patent a spinning frame machine that was able to produce twisted threads by using wooden and metal cylinders rather than human fingers. That invention required fewer skilled workers and overnight markedly increased production output of woven fabrics. Unfortunately, the loss of jobs due to Arkwright's stock frame incited anger among textile workers, who regularly began to destroy textile machines. In 1779, Ned Ludd, a weaver from Anstey, near Leicester, England, broke two stocking frames in a fit of rage. Yet, despite the mayhem that the laborsaving devices were stirring up, the demand for a cheaper and faster means of production ultimately became the order of the day. Arkwright's invention was a harbinger of things to come. Even more technological advances linking one generation to the next followed.

By the late eighteenth and early nineteenth centuries, British cotton processing had advanced so far that British manufacturing methods had been universally adopted.

It would seem logical that, by the twenty-first century, society would have embraced the marvelous advances that have been continuously appearing. Yet, although each of the former technological revolutions yielded indisputable advantages, the new was and still is met with resistance.

In 2016, I was a solutions architect for a portfolio of twenty different aircraft systems. Each system entailed hundreds, sometimes thousands, of global users, and I was directly responsible for the day-to-day operations of two of those systems. It was my responsibility to enhance the performance of those two systems by improving their infrastructure, security, auditing, and so on. One of my least favorite tasks was code deployment, which is a process of replacing an old code with a new one. For instance, when individuals navigate to Amazon.com, whatever they see within Amazon.com is considered a code that is running in a production environment. It is visible to all users.

In most cases, developers build their code within a development environment (which is a mirror of a production environment) to test the code before it gets pushed to production. If, for example, Amazon.com wants to add another feature to its home page, such as a new tab called "My Preferences," then the old code has to be replaced by the one with the new feature. Unfortunately, this is not just a matter of deleting the old and replacing it with the new. It is a complicated procedure that involves a small cadre of programmers who have to test the code on the date of deployment. Each individual involved performs multiple steps to implement the entire update, and the complexity of each project varies according to the size of the system.

The first step is to make a copy of the old code as a backup, in case the new code fails. Then the entire system needs to be shut down while the old code is deleted, and a new code is uploaded. Next, new file permissions of the code files are installed. The new version of the system is started up, and the new code is given a trial run. Finally, the customer base is notified of the newly added features.

Having so many different steps, including the copying and moving of files, greatly increases the chances for human error.

It used to be standard practice that a code deployment was scheduled on a Friday night to cause the least disruption among users, since the majority of them, such as engineers, were primarily located within the United States, and most Americans don't work on weekends.

The code deployment process that my team and I were involved with usually took three to four hours. Even in the best-case scenarios, the time factor varied from one system to another. If, while testing the code in production, we encountered any problems caused by a bug or human error, it would sometimes take hours to troubleshoot and fix it.

In one instance, all in the name of code deployment, we arrived at work at 4:00 P.M. on a Friday and left at 3:00 A.M. the following Saturday. On Monday, after the team got back to work, bragging about how hard the deployment had been, we were heartily thanked by our manager.

Fifteen years ago, this approach was considered the best and most cost-effective. But in the past seven to eight years, most companies have automated the process. New platforms have been designed, which now automate the manual steps of shutting down the former system, copying its files back and forth, and relaunching the new system. I might add that all this is done with zero errors.

It now takes only a minute or two to complete an entire deployment by merely writing a script in a platform that is designed to automate the process. I only need to write a script one time and then schedule it to run on a specific date and time of my choosing.

When the automation of the deployment—or as it officially came to be known, "continuous integration"—became popular, I was eager to apply it.

It occurred to me that this timesaving method was going to make my life much easier. There would be no more coming to work late on Friday and staying until Saturday morning just for a code deployment! The script would free me up to concentrate on more important projects. Not only would I be relieved from a ton of minutia, but the company also stood to save significantly from the reduction in overhead costs.

To my mind, this was a win-win situation, but my teammates did not see it that way. Instead, most of them saw it as a threat instead of an opportunity, since many of them had been at their jobs for fifteen to twenty years, and some were even close to retirement. None of them said so to my face, but I could sense their growing anxiety. Giving up tasks that had once been part of their expertise was not easy for them, and I began to notice distractions popping up before the implementation of new projects.

One example of someone who dragged his feet was Don, who was a program analyst responsible for the code deployment of one of the systems in my portfolio, who also happened to be nearing his retirement. A pattern began whenever I needed information about the system Don was supporting. More than once, he emphasized that he was too busy to meet with me, and he would push the meeting a week back.

When he did finally show up for our meeting, he was grumpy and took every opportunity to stress how complex the code deployment was for his system and how any error by the script could impact the system's performance. I attempted to allay his concerns by reminding him that the same technology was now integral to Google, Amazon, and Microsoft, but my arguments fell on deaf ears.

But Don was not the only member of my team who resisted the automation. Also, to my surprise, was the lukewarm response of Liz, who was the project manager for another system in my portfolio. Regarding the new project, she said, "This is a good idea, but we need to test it first, and right now is not a good time because we're too busy working on business requirements. That's our top priority. Maybe we can work on it next year."

Even when I discussed the project with my manager, who just happened to have recently celebrated his fortieth anniversary with the company, I met some resistance. He seemed to be reassuring when he said, "Let's keep trying to find a way to implement this project. It seems like a very good idea!" But then he, too, placed it on a back burner.

I felt as though I were falling down a rabbit hole because it appeared that no one but me was seeing the value of implementing the project.

The negative responses were not because continuous integration was not advantageous. Rather, the opposition stemmed from the basic human resistance to change. Letting go of the familiar and foregoing one's comfort zone will always be challenging, since it can bring up painful feelings. Most of us are threatened by the notion that we are not enough and that something faster, more efficient, more cost-effective, and more profitable will show us up for being less capable and less necessary than we supposed.

There used to be a radio program that always had a quote for the day. On one occasion, the quote, which was both humorous and undeniable, was, "Cemeteries are filled with indispensable people."

That taught me a good lesson because I believed that we should always be expanding our knowledge base and never get too used to a comfort zone. Nevertheless, I am totally aware that technical know-how is always racing down a conveyer belt, but falling off and being replaced with ever-increasing speed.

Although some people regard returning to school or taking a course as proof of their inadequacy, I consider it stimulating.

Whenever I encounter obstacles, I do my best to see them as opportunities for growth. I say I do my best because, like everyone else, I have to make a conscious effort to "reframe" progress in positive terms. Putting things in a better context can help to take out the sting.

But to continue with my story, I only made minor progress in applying continuous integration within non-production environments, and with only *some* of the systems at that. Later, when an opportunity presented itself, I gladly accepted an offer to take an entirely different job within the same company. It was time to move on. I was more than ready to work on a fresh challenging project that would also help to further my skill set.

Fast-forwarding four years, I heard through the grapevine that the company had retired most of the systems I have discussed here, because, having become obsolete, they were now too costly to maintain. As a result, many of my previous coworkers got laid off. That made me wonder: *What if they had decided to be part of the continuous integration project by going with it instead of against it?*

One thing, however, was undeniable: accepting the new direction things were moving toward instead of resisting them would have, at the very least, given my teammates a new skill set for their job search.

The Second Industrial Revolution

Some historians argue that the First Industrial Revolution was in its twilight by 1850. Britain's economic miracle had slowed down as global integration of the empire's advances had by now successfully re-invigorated economies around the world. The various new inventions that followed failed to revolutionize production output. Worldwide production had stayed on a continuum from the original wave of industrialization.

Other scholars contend that the Second Industrial Revolution began around 1850 and lasted up until 1914, with the start of World War I. Some historians have even estimated that that revolution lasted until as recently as 1950. Early on, the new revolution could no longer be labeled British, because it now included every industrialized nation of the late nineteenth century, with all of them contributing to the upcoming changes (History, 2020).

Some of the most important inventions in the Second Industrial Revolution took place in metallurgy. The steelmaking process, invented by Henry Bessemer, made a huge contribution. Known as the "Bessemer Process," it was the first inexpensive method to mass produce steel. Bessemer realized that impurities could be removed from steel by oxidation, the blowing of air through the molten iron. Steel production quickly became the backbone for new construction and designs in various fields. With the newfound alloy in demand, it became obvious that producing steel at the rate needed would require dynamic new energy sources. The power of water or steam was now inadequate.

Scientists like Michael Faraday, Alessandro Volta, Hans Christian Ørsted, André-Marie Ampère, and Georg Ohm were among the leaders who explained the scientific basis of electricity. By 1821, Faraday managed to create the first electric motor, which demonstrated what electricity was capable of. The concept of electricity had been almost purely theoretical until the late nineteenth century, when electric current was finally harnessed by engineers and by inventors such as Thomas Edison.

It was during the Second Industrial Revolution that the

introduction of electricity gave rise to fabulous possibilities for the new source of energy. Its almost limitless applications helped to forge inventions that people had only dreamed of before.

In 1886, the first functioning and commercially available motor vehicle was patented by Carl Benz, a German engine designer and automotive engineer. The spawning automotive industry could now be found in France, Britain, and the United States.

Initially, engines were powered by coal, gas, or gasoline. For millennia, crude oil had been used for lighting, heating, and tar, but its industrial uses were not discovered until 1847 by Scottish chemist James Young, whose distillate technique converted petroleum into paraffin. The resulting light oil became ubiquitous in everyday life—used for lighting, heating, cooking, and even machine lubrication. However, with the explosive growth of the automotive industry, petroleum production became a gargantuan industry.

Miracles of the age continued to dazzle and amaze humanity. These included telephones, airplanes, cameras, sound recorders, motion pictures, nuclear fission, and new vaccines that vanquished life-threatening diseases.

The Third Industrial Revolution

Automating the Manufacturing Process

Until 1968, the only way to control machinery was by using relays, which worked by utilizing a coil that was energized or de-energized by a magnetic force. When energized, a relay pulled a switch to the "on" position. When de-energized, a relay returned the device to the "off" position.

Imagine a machine that has a motor that you want to control with an "on" or "off" switch. Formerly, the only way to run machinery was via the relay between the power source and the motor. Someone would control the motor by either energizing or de-energizing its relay. This type of relay is known as a "power relay."

Now imagine a textile mill operating several machines simultaneously, such as a woolen mill machine, a thread-winding apparatus, bleaching and dyeing equipment, and so on, with all of them controlled by power relays. Not long ago, textile mills had to make room for electrical cabinets, which were filled with power relays (Green, 2021).

Source: https://commons.wikimedia.org

Naturally, the number of relays increased as businesses grew. The huge bulk of relays required continuous organization for each specific production process. In addition, if one relay went down, troubleshooting could sometimes take up hours of valuable time. Such delays created domino effects, which then caused havoc with production schedules. That, in turn, required intensive labor to continuously operate and maintain giant networks of relays and their corresponding machines.

By the 1960s, the term *computer control* had become part of the lexicon of large corporations such as General Motors and the Ford Motor Company. According to Dick Morley, the undisputed father of the Programmable Logic Controller (PLC), who worked for General Motors, "The programmable controller was detailed on New Year's Day, 1968" (Green, 2021). The PLC was initially designed to replace a factory's banks of relays with a single device, which was easier to maintain and could be programmed with the already accepted relay ladder logic.

However, there were inherent drawbacks to PLCs. The first PLC was large, prohibitively expensive, and initially incompatible with factories' infrastructures. In time, though, the core functionality and simplicity of PLCs made them indispensible for any factory. Keep in mind, however, that today's PLCs are far smaller, more powerful, and faster than their predecessors.

With transistors, integrated circuits, and computers now on the scene, the Third Industrial Revolution took hold. Those new technological marvels reduced effort, increased velocity, enhanced accuracy, and automated vast numbers of factory jobs. Now there was a frenetic pace of production lines across multiple geographical locations, which led to the formation of supply chain management, enterprise resource planning, inventory management, shipping logistics, product-flow scheduling, and assiduous detail tracking.

Like something out of an Isaac Asimov science fiction novel, electronic and information technology industries were jettisoning outdated modalities at a warp speed. Worldwide digitization began with the use of semiconductors and mainframe computers in the 1970s, and developed further with personal computers in the 1980s. In the 1990s, the floodgates opened wide with the introduction of the Internet, followed by smartphones in the 2000s. All of these accelerated movements and exchanges, along with the quantum leaps in telecommunications technology, brought about the globalization of our planet. That metamorphosis created an altogether new economic narrative that has impelled businesses around the world to radically transform themselves by outsourcing their human labor costs overseas.

Enterprise Architecture (EA)

During the Third Industrial Revolution, when personal computers started to become popular, new demands for software arose. Computers began to be used to write software systems to increase organizational productivity and efficiency, especially in areas like product lifecycle management, enterprise resource planning, payroll, human resources, and supply chains.

However, when business managers began to undergo extensive organizational changes, these software systems became obsolete, which created a new predicament. With so much major redevelopment under way, newer software programs were needed. But there were no guides who could explain how to build such software, especially on an enterprise level, or how components of such enterprises could be integrated to work together.

In 1980, John Zachman, an IBM engineer, published a paper that differentiated "organizational responsibilities" and "processes" as two different phenomena. Zachman believed that systems should be designed to automate processes and not to encode organizational responsibilities, because processes and organizations frequently change independently of each other (Zachman, 2008). In 1987, he launched his framework for how to build a business and systems architecture for enterprises.

At around the same time, the U.S. Department of Defense (DOD) realized its need for a framework that could provide enterprise-level guidance to identify services, standards, concepts, components, and configurations. Such guidance was instrumental in the development of a technical architecture that could meet the DOD's missions. In 1986, the DOD started to develop the Technical Architectural Framework for Information Management (TAFIM), whose first draft was completed in 1991. In 1995, the DOD made the framework available to a global industry consortium called "The Open Group," which further developed the framework and made it available to the public under a free license. The Open Group published the first enhanced version of the framework under a new name: The Open Group Architecture Framework (TOGAF). Since then, TOGAF has become the most common framework to run enterprise architecture.

While project management is the discipline of initiating, planning, executing, controlling, monitoring, and closing projects, enterprise architecture is the process of translating business strategies into an enterprise-wide set of capabilities, which may include all of an enterprise's information, technological services, processes, people, and infrastructure. Each capability can help with the current and future objectives of the enterprise. However, the enterprise is looked at as a collection of organizations that share a common set of goals. Based on that definition, an enterprise can be a governmental agency, a corporation, or a chain of geographically distant organizations that are linked together by common ownership. For instance, a government can be regarded as an enterprise, and cities can be regarded as organizations that belong to that enterprise.

Since the 1990s, many enterprises have adopted an enterprise architectural practice as the best method for determining how an "organization" can effectively achieve its current and future objectives. Frameworks like TOGAF were created to give organizations the ability to plan, analyze, design, and ultimately implement initiatives such as a "smart city"—a concept to be discussed at length in Part 2 of this book. TOGAF can also help an enterprise to achieve digital transformation, as discussed in Chapter 1. According to Gartner, companies with a strong enterprise architectural foundation have 25 percent lower information technology costs, which plays a central role in reducing the complexity associated with digital transformations.

The Fourth Industrial Revolution

In economics, the term *fixed cost* refers to a cost factor that does not change with increases or decreases in the number of goods or services that are produced or sold. However, the term *marginal cost* refers to the change in the "fixed cost" of producing additional units of goods or services. Enterprises today are digitally transforming their infrastructures and business models to increase their production, reduce marginal costs, win more consumers, and stay in the black.

Economists have long had the goal of bringing marginal costs down to near zero. With the technological revolution of these past few years, a new term has been coined: *extreme productivity*. Extreme productivity has made it possible to bring marginal values down to zero, which has made communication, energy, transportation, and countless physical goods and services very cheap.

The term *prosumers* refers to individuals or entities that are both consumers and producers—for example, someone or some organization that sells books, music, or videos on Amazon, or someone who rents his or her home out on Airbnb. In recent decades, millions of consumers have become prosumers, bringing their marginal costs near to zero. The emergence of this new demographic has greatly impacted many industries—in some cases, putting them out of business. Two prime examples of this would be traditional taxicab services versus Uber, and the analogue-based Kodak Company versus the digitally based Sony Corporation. Bringing about "extreme productivity" has benefited both consumers and producers on different levels. "Information goods" have become today's commodities, helping us to transition away from a brick-and-mortar economy.

The radical changes and rapid transformation of the way businesses are being conducted today are impacting organizational structures, jobs, wages, and social behavior. The very nature of work has changed. The world is entering a new chapter in human development, enabled by extraordinary technological advances equivalent to those of the First, Second, and Third Industrial Revolutions. Aptly, this current revolution is called the Fourth Industrial Revolution (Industry 4.0).

In 2013, represented by both the Federal Ministry of Education and Research and the Ministry for Economic Affairs and Energy, the German government first mentioned the term *Industry 4.0*, thereby creating a new national strategic initiative. The dynamic strategy was meant to drive digital manufacturing forward by stepping up the processes involved in digital transformation and the connection between products' value chains and their business models.

The initiative was also aimed at supporting research and implementing standardization across the whole country. The German government was one of the first to fully concede that the world was entering a new era. By strategizing the breakthrough initiative, it acknowledged that the world would never be the same again. The initiative was more than a new strategy; it was a call to action. Today's ethos is that one either becomes proactive and aligns with the forward momentum of Industry 4.0 or is left behind. In 2016, Klaus Schwab, a German professor and the chairman of the World Economic Forum, published a book defining Industry 4.0 (Schwab, 2017), which divided the Fourth Industrial Revolution into three components: physical, digital, and biological.

THE PHYSICAL COMPONENT includes autonomous vehicles, three-dimensional printing, advanced robotics, and new materials.

Autonomous vehicles include driverless cars, trucks, buses, tractors, drones, aircraft, boats, and so on.

3D printing is the process of manufacturing any physical object by creating it layer by layer, thereby forming a digital 3D drawing or model. This process can enable the creation of very complex shapes, while using few materials and little time. The materials utilized for 3D printing can be as diverse as plastic, metal, powder, or liquid. This technology can materialize anything from clothing to footwear, to food, to machine parts, to human organs. In some cases, a robot can operate a 3D printer, making it capable of not only autonomously constructing other machines from scratch, but also even recreating itself.

Advanced Robotics: During Industry 3.0, robotics were mostly confined to tasks in specific industries, such as automotive or aerospace factories. Nowadays, however, robots are increasingly used in medical procedures, agriculture, military operations, food preparation, and so on.

New materials include self-healing substances that imitate a living organism's ability to heal itself after an injury. These smart materials, which are light, strong, recyclable, and adaptive, are commonly found in sensors, actuators, and artificial muscles (particularly as electroactive polymers for prosthetics).

THE DIGITAL COMPONENT includes the Internet of Things (IoT), which is one of the main concepts that enable Industry 4.0. The IoT, which is the bridge between the physical and the digital component, refers to a system of interrelated things, which can be products, machines, animals, people, or any physical object. For instance, a "thing" can be a person with a heart monitor implant, a farm animal with a biochip transponder, or a car that has built-in sensors to alert a driver about nearby vehicles. The "thing" can be any natural or man-made object that can be linked to the Internet and encompass the transference of data.

THE BIOLOGICAL COMPONENT includes genetics and synthetic biology.

Genetics: In 1990, the U.S. Congress established funding for the Human Genome Project, setting a target completion date of 2005. The project's goal was to place, in a completely accurate sequence, the three billion nucleotide base pairs within the human genome. In addition, the resulting mapping and identification of the genes present in human DNA needed to be stored in a sequential order within the database. The project concluded in 2003 for $2.7 billion. Today, a genome can be sequenced in a few hours for less than $1,000. Scientists are now able to test how specific genetic variations generate particular traits and diseases without having to resort to trial and error.

Synthetic Biology is the ability to redesign and customize organisms by writing DNA to produce substances. A good example would be creating microbes that seek and destroy tumors in the body before the body self-destructs. Synthetic biology aims to make biology easier to engineer. The fusion of chemistry, biology, computer science, and engineering enables us to go from idea to product faster, less expensively, and with greater precision than ever before. According to biospace.com, the synthetic biology market was valued in 2018 at $4.96 billion and is projected to expand at a compound annual growth rate of 26.3 percent between 2019 and 2027 (Bhisey, 2020).

Smart Factories and Their Digital Twins

In Industry 4.0, machines no longer need to be run by humans. Nowadays, they can sense, think, and make decisions on their own without human intervention. Artificial intelligence has leveraged Industry 4.0's digital capabilities.

Today, a typical factory consists of many machines and various production components, each of which entails unique specifications. Over time, the condition of the machines can degrade, thereby producing slight variations in the products. In a traditional manufacturing setting, these variations often went undetected.

By contrast, a machine in Industry 4.0 is equipped with a multitude of Internet of Things sensors, each of which generates real-time data about the machine's health status in regard, for example, to heat, pressure, or error warnings. The data can be transferred via the Internet to remote servers and then processed and made visible to workers and engineers. The beauty of this architecture is that no staff members need to be near the machines. To further ensure against any errant malfunctions, machines now have “digital twins,” which can utilize historical and real-time data to think, predict, and react just like the human cortex, the area of the brain known as the seat of intelligence.

Source: https://witanworld.com

One of the major challenges that all factories want to avoid is the failure of their operating equipment. Equipment failure can impact an entire industrial unit. For instance, it could mean that customers do not receive their orders on time, or workers lose compensation, or managers have to take the time to submit reports that explain what happened to account for production losses.

However, a digital twin has all of the necessary data, artificial intelligence, and machine learning algorithms to prevent any unscheduled or unplanned downtime by sensing and predicting the problem before it happens. This marvel of today’s technology offers any kind of production equipment numerous benefits that translate into significant cost savings and increased revenues for business enterprises. Listed below are some examples of the tasks for scheduled maintenance that a digital twin can predict:

- Oil changes
- Filter changes
- Equipment lubrication
- Cleaning
- Calibration
- Visual inspection of components for possible replacement
- Control optimization

A typical factory will have multiple machines. Each of those machines, in turn, will have its own digital twin, and those twins will integrate and communicate with each other via a "digital thread" of data that facilitates the communication of data across a product's entire lifecycle. A factory in the Industry 4.0 world is identified as a "smart factory" due to its ability to use and integrate data generated from machines, devices, systems, and workers. Any smart factory can self-optimize performance across a broad network, self-adapt to and learn from new conditions in real or near-real time, and autonomously run entire production processes.

Industry 4.0 will touch every aspect of our lives, and the word *smart* will be attached to a variety of businesses and occupations. We will see smart factories, smart cities, smart homes, smart farms, smart buildings, smart military bases, smart cars, and so on.

In Chapter 1, I detailed how many companies and enterprises around the world are currently going through a massive digital transformation, changing the way they operate to keep pace with the times. Depending on the size of the company, such a transition can grow into an enormous task. The undertaking of such a massive effort is the result of many enterprises' survival instincts kicking in. However, some companies will be unable to meet the challenging transition to the world of Industry 4.0.

People are also rapidly changing, but I am not sure how many of us have noticed that. For instance, in August 2019, I was invited to a wedding in Lithuania by Katherine, a friend of mine of Lithuanian descent. She and I planned the following itinerary: first, land in Vilnius, the capital; then rent a car and head to the wedding; and finally, drive to Katherine's birth city of Klaipeda, about two hours' driving time from Vilnius.

On the morning after the wedding, Katherine and I set off for Klaipeda. Once we were on the road, the beautiful scenery delighted me. The bucolic landscape included farmers driving their tractors alongside grazing cows and horses. The road was almost devoid of any other cars. Occasionally, we saw a gas station, or another motorist would pass by, but nothing else.

The speed limit was 90 kilometers, or about 55 miles, per hour. Now and then, we passed signposts marking the speed limit. While enjoying the scenery and talking to my friend, I couldn't resist the temptation to increase my speed to 70 miles per hour, which is the standard speed limit on most U.S. highways, and which I find most comfortable. When I stepped down on the accelerator, I rationalized that I could easily see any police cars ahead or behind me on the straight empty highway.

After driving for forty-five minutes, we got a phone call from the rental car company. Although neither of us could tell whether the voice was human or automated, it informed Katherine that she had been driving over the speed limit for the past fifteen minutes! It addressed Katherine as the driver because she had made the reservation in her name.

Our location was identified at exactly the place where

we had exceeded the speed limit, and we were warned that if we did not slow down, the car engine would shut off, which shocked Katherine!

I said with a smile, "The car is not just a physical object anymore."

Then I pulled over to the side and handed the keys to Katherine, so that she could take over the driving.

The digital twin of our car, thinking that we were putting ourselves in danger, made a rational decision to warn us. If we had resisted the warning, we would have been stranded in the middle of nowhere. All of that had been done without any human intervention.

Ultimately, this smart architecture leads to fewer traffic accidents, fewer policemen, fewer traffic courts and judges, and a better experience for people, because no one likes to be chased by a police car with flashing lights. And certainly, no one likes to get a ticket and pay a fine.

Such an architecture also reduces the possibility of police bias against drivers of a certain race, age, gender, and so on. The picture below came from Statista, a German company that specializes in market and consumer data. The graph clearly illustrates that, after the smart architecture was introduced, traffic incidents in Lithuania went down almost 80 percent. Keep in mind that Lithuania, like Germany and many other countries, joined in the Industry 4.0 race by launching its own nationwide strategy.

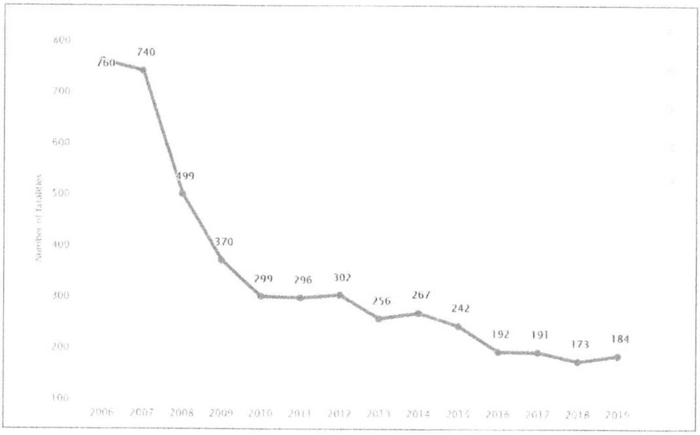

Source: https://www.statista.com

Wouldn't we all like to see this result in our own country? We must ask ourselves why our states and cities are not utilizing similar technologies to provide better living and safety standards for our citizens. In Part 2, I will examine some of the challenges that hinder our states and cities from digitally transforming their businesses to Industry 4.0 standards.

Chapter 3
Agriculture in the World of Industry 4.0

Usually, when one thinks of the word agriculture, that does not bring to mind technology. Not only have the combined powers of the Internet of Things and artificial intelligence begun to create almost limitless opportunities to serve and influence humanity at large, but they have also specifically extended their influence into agriculture.

Today, there are over one billion more people on Earth than there were twenty years ago. The total number of humans inhabiting our planet now exceeds seven billion. With this increase in population, the ratio between resources and number of mouths to feed has reached a tipping point. Henceforth, our traditional methods of agriculture will prove to be woefully inadequate.

Let us ponder a few interesting points. To begin with, it is predicted that 70 percent of the world's population will inhabit urban areas by 2050. People living in urban areas care more today about the quality and health of their food than they used to (Ayaz et al., 2019). That means that, in addition to the growth of the population, there will be a proportional increase in the demand for more goods, and in turn a greater demand for the high quality of those goods. Agriculture, in addition to providing food, systemically provides a wide range of other necessary goods, including beauty products, cotton, rubber, and so forth. All of these products and by-products will continuously draw upon our finite resources, which is a compelling reason to implement Agriculture 4.0 as soon as possible.

Only certain pieces of land on our planet are suitable for growing produce. Statistics show that land utilized for food production has experienced a continuous decline since the end of the nineteenth century (Bienhaus, 2017). It follows from this that optimizing the use of land is vitally important.

As demands rise, we will need ever-higher quality, productivity, efficiency, and sustainability of all our agricultural goods. Although the term smart agriculture in the world of Industry 4.0 is used interchangeably with smart farming and precision agriculture, the new revolution in the agricultural industry is not driven solely by technological advancements, but also by other factors. For example, low-cost labor could easily be undermined as the result of a political decision.

In the past few years, agriculture has transformed from small and medium farming operations to highly industrialized and commercial enterprises. It is now on a technological par with other industries. Not unlike its manufacturing counterparts, agriculture now requires measurements, data, and controls that can be automated digitally for planning and management. Every component is integral to maintaining the balance between costs and production.

We are witnessing the dawn of a new era in agriculture. With the appearance of agricultural robots and their digital twins, husbandry has become more sophisticated than we ever dreamed of. Such pieces of equipment are now performing a range of activities such as planting, watering, weeding, fertilizing, spraying, thinning, picking, packing, and transporting. These technologies are providing us with solutions to many age-old quandaries, such as how to respond to drought, how to optimize crop yields, how to irrigate land, how to control pests, and so on.

The escalating population has also created a greater demand for fresh water. The growing number of global disasters being brought on by climate change, along with the planet's population reaching a critical mass, has made the life-sustaining resource of water more precious than ever. Fortunately, Industry 4.0 technologies such as the Internet of Things, big data, machine learning algorithms, and artificial intelligence can help to alleviate farmers' struggles in managing water usage. Optimizing the amount of water needed to produce the highest yields prevents excess water usage and malnourished crops. Yet, despite the almost indisputable benefits that precision agriculture brings, it still faces opposition.

Brazilian researchers at Centro University studied the applications used in precision agriculture (Monteleone et al., 2019), which brought about the creation of the Smart Water Management Platform, or SWAMP. The researchers tested a smart agricultural system in four places: two in Brazil, one in Italy, and one in Spain. In Brazil, soybeans and cotton were grown on one farm, and grapes for wine were grown on another. In Italy, smart water distribution was used for a vineyard, as well as for fields with pears, apples, soybeans, and corn. In Spain, smart water distribution was applied to an intercrop irrigation system that included bay leaves chicory, herbs, lettuce, and spinach.

In all three countries, the system was divided into five layers:

Devices and Communication: a variety of Internet of Things sensors and actuators to acquire solid data regarding weather, plants, and metrics.

Data Acquisition, Security, and Management: the protocols for collecting, analyzing, and maintaining data.

Data Management: software components responsible for

storing, processing, and distributing data.

Water Distribution and Irrigation Models: traditional agricultural models for estimating plant water requirements, by way of drone-generated images as well as sensors that detect soil moisture.

Water Application Services: irrigation services that provide user experiences for farmers and water distributors.

These technologies now provide real-time assistance in decision-making. They make it possible for a more efficient system of distribution and for a better management of resources, such as scheduling irrigation, choosing which crops to cultivate, choosing which machines to use, and knowing how and when to complete such operations.

These technologies are not for the future. They are already here. These smart systems are expected to be adopted soon by farmers who are still running their farming businesses by the time-honored method of observing and learning through old-fashioned hands-on experience.

A recent article on the website of The National Institute of Food and Agriculture (n.d.) stated: "Farmers require ongoing education to stay aware of fast-moving developments in technology, science, business management, and an array of other skills and fields that affect agricultural operations."

Agriculture in the world of Industry 4.0 is a digital process-based system that involves precision methods, utilizing the Internet of Things, machine-learning algorithms, and cyber-physical systems incorporated within machines and equipment. All of these components have helped to improve decision-making and enhance productivity. Farmers today are facing unique challenges in catching up with and adapting to the massive technological advances in front of them, which require specialized education and training on a governmental level to ensure success. The integration of these advantages into day-to-day operations will lead to:

- Less harm to the environment
- Reduction of food contamination
- Reduction of the need for water and chemicals for crops
- Increased profits

According to the Bureau of the Census, the American Community Survey, and other federal statistical sources, the overall educational level of people in rural areas has increased in the past few decades. However, it is still lower than the level in urban areas, especially for individuals with a bachelor's degree. In 1960, 60 percent of the rural population, aged 25 and over, had not completed high school. By 2018, that figure dropped to 13 percent. Over the same fifty-eight years, the number of adults in rural areas, aged 25 and older, who held a bachelor's degree or higher, increased from 5 percent to 20 percent. In urban areas, that last figure stood at 35 percent by 2018. The proportion of adults in rural districts with a bachelor's degree or higher increased by 5 percent between 2000 and 2018. Rural adults not holding a high school degree or GED certificate declined by 11 percent (Farmer, n.d.).

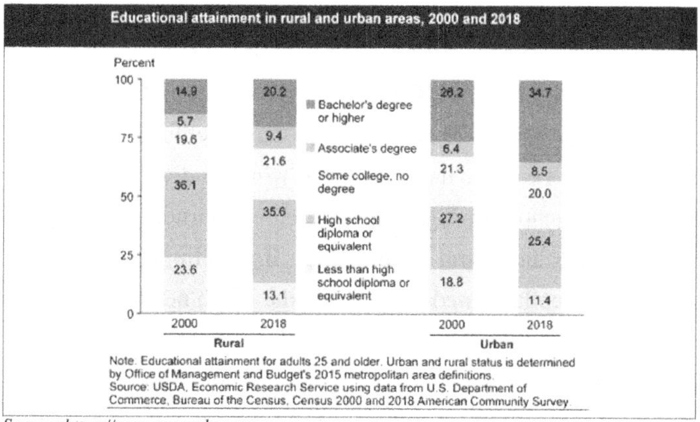

Source: https://www.ers.usda.gov

I believe that education will be the most significant factor in determining any nation's ability to adapt to Industry 4.0, regardless of its urban or rural demographics. Ideally, the level of education will need to be approximately equal between individuals who work on farms and those involved in the global corporate complex. That being said, there should be more collaborative efforts between governmental agencies and farmers.

According to a study conducted by the U.S. Federal Communications Commission, an alarming number of rural regions throughout the United States still do not have access to the Internet. By contrast, in urban areas, 97 percent of Americans have access to high-speed Internet services. In rural areas, that number drops to only 65 percent. On Tribal lands, barely 60 percent have access. All told, nearly 30 million Americans cannot reap the benefits of the digital age (National Geographic Society, n.d.).

There is no doubt that Internet access is fundamental to the new era of agriculture. Without it, farmers will be forced to rely on outdated technological agricultural methods, which will cause still further shortages of agricultural products on a national scale. At present, we don't have enough food and other agricultural products to keep pace with the swelling urban population.

As of today, there is also a huge gap between rural versus urban incomes. The table below illustrates the difference in earnings based on the census data of 2019, in which individuals holding the same educational degree in urban areas earned much higher incomes than their rural counterparts. If governmental agencies and farmers collaborated to facilitate the much needed transition to smart agriculture, these discrepancies would be leveled out. Farmers becoming involved now in smart agriculture will need advanced technological skills to run the automated software and machines required in the near future. The next generation will see farmers in the field who hold associate, bachelor, and graduate degrees.

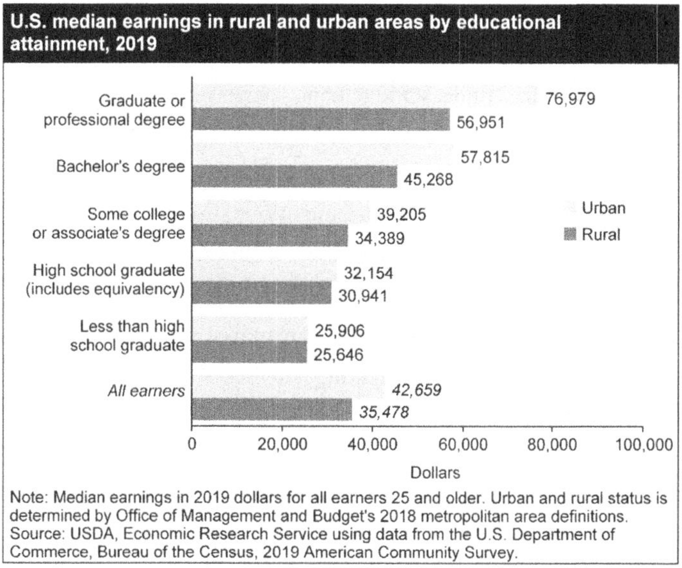

Source: https://www.ers.usda.gov/

I hope that smart farming will bridge the gap between urban and rural incomes and lead to greater equality, both in the job market and in overall standard of living.

Throughout my career in the tech industry, I have met many coworkers from states like Oklahoma, South Carolina, Ohio, Arizona, and Texas. A lot of them grew up on a farm and later moved to a city to pursue dream jobs that invariably dealt with much of the latest technologies.

When a number of them described country life to me, they said they longed to return to it. Smart farming makes it possible for people to work in highly sophisticated tech jobs while living in rustic environments.

Contrary to the popular expression, the grass is not always greener on the other side of the fence. The forward thrust of advancing technologies has an ironic resonance. By availing ourselves of the many benefits of smart agriculture, we are simultaneously displacing many individuals from their livelihoods. A great many jobs currently performed by men and women are being transformed from their functions as human endeavors into autonomous and tireless movements performed by robots and software programs. In the very near future, an entire farm can and will be run by a digital twin, which will mean that a significant drop-off in manual labor will occur. Whatever jobs remain will be rewarded only to those with higher education in agriculture, especially those specializing in software engineering. This is a cogent argument for why a governmental strategy needs to be implemented to prepare unwitting citizens for these imminent changes.

Not to put too fine a point on it, but I must again stress that a digital infrastructure is vital for a smooth transition to Industry 4.0. As I mentioned in Chapter 1, responsible and capable leadership needs to be at the helm to create the proper culture for the digital transformation to succeed.

Many companies and organizations are currently setting in motion initiatives for digital transformation. Likewise, our governments should emulate the private sector's strategic advances, not only to get on board with current technology, but also to prepare the public to cope with the vast transitions taking place.

In Chapter 2, I highlighted how Germany in 2013 was the first country to officially launch such a strategic direction. After that, many countries followed Germany's lead. For instance, in 2018, the Indonesian government launched "Making Indonesia 4.0," a strategy that is helping to propel Indonesia into the ranks of one of the top ten nations in gross domestic product by 2030. That strategy includes tax deductions for companies that are actively conducting and investing in research related to Industry 4.0. The plan includes incentives for companies that are sponsoring their employees to pursue specialized education to prepare them to cope with new technologies. In a recent study published by McKinsey, it was estimated that digital transformation could add $120 billion to Indonesia's economic output by 2025. More than a quarter of that figure, about $34 billion, would be from manufacturing (Agarwal et al., 2020).

So, what is the Industry 4.0 strategy in the United States? In Part 2, I will share some examples of how attempts at digital transformation have fallen to the wayside or ended up as wasted efforts, entirely due to a lack of strategic direction, knowledge, or leadership.

Part 2

A Better Standard of Living Through Technology

In this part of the book, I will detail the current lag of our cities in keeping up with all of the ongoing changes associated with Industry 4.0. In my efforts to engage with the officials of my own community of Long Beach, California, I have encountered both reluctance and a steadfast refusal to participate in any dialogue that involves the changes needed. Admittedly, I came out of these encounters somewhat disheartened, but at the same time, much more enlightened.

With everything about us changing with such rapid-fire succession, no part of our society can afford the luxury of lingering. I, for one, believe that we must entirely rethink our understanding of "community." Residents of today's communities should be seen as "shareholders," to use a corporate term. The objective of corporate shareholders is to influence the governance of a firm to meet its goals. Similarly, residents should effectively influence the governance of the communities they have invested their lives in.

Passively allowing the people in charge to make arbitrary changes without the knowledge of their constituents is no longer a tenable proposition. The mercurial advances being made by the many technologies emerging within Industry 4.0 demand that we revise our complete cultural environment. Change is required in the way we not only conduct businesses, but also in the very leadership that govern our communities. We must have a new digital leadership style that embraces change and encourages citizens' involvement with their communities.

Chapter 4
The Smart City

Cities Now and Then

A pólis or city-state was a community structure conceived of by the ancient Greeks. City-states, or póleis, were where the majority of the population lived. The government buildings and temples in the city-states were usually positioned at the top of a hill, or acropolis. This focal point served as the center of trade, commerce, culture, and political activity. Only a few city-states reached populations that exceeded a thousand persons. The main city-states of ancient Greece were Athens, Sparta, Corinth, Thebes, Syracuse, Aegina, Rhodes, Argos, and Eretria (National Geographic Society, n.d.).

By 1950, the total urban population of the planet reached 746 million, and has since risen to 3.9 billion. Metropolitan populations are expected to reach 6.4 billion by 2050 (Herald Globe, 2014).

Municipal governments have undergone many changes over the centuries. In the United States, the concept of local government has been deeply rooted in the history of the nation, with citizens historically maintaining strong support for their local governments. Yet, despite that support, most Americans are ignorant of exactly what their local governments do, how they are funded, and to what degree they impact daily life.

In the United States, local governments are broken down into counties, municipalities (cities and towns), special districts, and school districts. Traditionally, these branches of government are more responsive to citizens than state or federal branches are. Cities have the power to tax their citizens to pay for city services. Municipalities of all sizes are responsible for providing police protection, water, sewage, street maintenance, and, in some cases, airports, zoos, sports stadiums, and libraries. Cities also have judicial court systems that facilitate cases involving the municipal laws adopted by each community. In addition to all of these facilities, municipalities are also responsible for providing any number of other services that can make life better for their residents.

Thirty years ago, people were not using the Internet as they do today. Communication was still mostly done by phone calls and face-to-face contact. Before the Internet, it was more difficult for a municipality to communicate with its residents. For example, unlike today, warning letters of any impending power cuts or available disaster relief had to be mailed to every household. In some cases, the radio and TV were also used to make public announcements.

Back then, if people wanted to file a report or pay for a traffic ticket, they routinely made trips to the city hall or their local police station. Those days are behind us now. Thanks to the development of technologies and the Internet, people can now file reports or pay for traffic tickets online. Today, communities can issue alert messages to their residents' phones without delay. But is that enough?

What Makes a Smart City?

Cities in the world of Industry 4.0 are being impacted in the same way as the private sector. Communities today are being challenged to maintain the knowledge base and leadership needed to keep pace with the advances that Industry 4.0 has brought on.

The term *smart city* has become a buzzword nowadays, but often mistakenly. Most communities are still not incorporating the technologies needed to redefine themselves as smart cities.

The definition of a smart city has changed over the past ten years. It is a concept that is still evolving, as is Industry 4.0. Below are recent definitions of a smart city from different sources:

- "A smart city uses information and communications technology (ICT) to enhance its livability, workability, and sustainability" (smartcitiescouncil, n.d.).

- "A city that uses information and communication technology to sense, analyze, and integrate the key information of core systems in running cities" (IBM, 2010).

- "A city connecting the physical infrastructure, the IT infrastructure, the social infrastructure, and the business infrastructure to leverage the collective intelligence of the city" (Harrison et al., 2010).

- "A city in which all of the conditions of all its critical infrastructures are monitored and integrated" (Hall et al., 2000).

- "A smart city is a city seeking to address public issues via information and communications technology–based solutions on the basis of a multi-stakeholder, municipally based partnership" (European Parliament, 2014).

There could be even more definitions that I may have missed. For now, though, my own working definition of a smart city is a community that is using many, if not all, of the current technologies available to provide a better living standard for its citizens. Such a community utilizes the various robotics, software, and data processing systems available today to reduce pollution, improve traffic flow, simplify parking utilization, advance health care, administer governance, and boost the local economy. However, for this to happen, the shareholders, who are the community's inhabitants, must engage with their local government, since their input and involvement are crucial to the success of initiating smart city initiatives.

According to the United Nations Population Division, 54 percent of the world's population is currently urban, and that will swell to 68 percent by 2050. Cities are engines of economic growth and are central to achieving global aspirations, such as the UN Sustainable Development Goals and Paris Agreement on Climate Change (Elmjid, 2017).

The U.N. Population Division expects that populations in urban areas will double in the foreseeable future. The great cities of today are quickly morphing into the "megacities" of tomorrow.

Smart agriculture will be key to maintaining the quality standards and affordability of food for such supercities. But food is not the only concern. Of equal importance is how cities will provide better living standards for their citizens to match that growth. Thus, it stands to reason that we must adopt the technologies available today to meet the challenges of tomorrow. Not enacting appropriate strategies today will further exacerbate coming future crises.

Consider some of the challenges that our cities are faced with today. In Los Angeles, for example, there is the issue of inadequate parking. In addition to that, there are now over 35,000,000 registered vehicles in Southern California, which has reduced the flow of rush hour traffic to a sluggish 15 miles per hour! The accumulated discharge of carbon monoxide from so many motor-driven vehicles continuously poisons the very air the public breathes. In addition, wildfires and mudslides lay waste to entire neighborhoods throughout the Los Angeles Basin. By 2050, these problems and others will increase in magnitude as the population continues to multiply. Traditional solutions such as adding more parking lots or increasing the width of streets will be untenable, since there will not be enough space left in the city for expansion.

The only way for Los Angeles to resolve its current and future urban problems will be to utilize the know-how that Industry 4.0 offers. While technologies like the Internet of Things, Big Data, machine learning, prediction algorithms, and digital twins will run all future smart cities, today's working rhythm and culture in our cities must change now. Our communities must embrace the ongoing digital transformation with the same initiative and enthusiasm as are being shown by the private sector.

Bureaucrats Can Undermine Smart Cities

My passion for the concept of smart cities started in 2018, while I was teaching a course on "Networking and Distributed

Systems" at California State University, Long Beach. To illustrate an example of a smart city, some of my students cited several initiatives adopted by the city of San Diego. After their interesting presentation, I was curious to know more about those initiatives, so I conducted my own research.

I soon learned that San Diego's Department of Sustainability and Environment had launched a program called the Climate Action Plan, whose goal was to eliminate no less than half of all greenhouse gas emissions within the city by 2035. The city also had an initiative that required all electricity to be generated from renewable sources by the same year. Both plans set forth common-sense strategies to attain reasonable targets, which included six factors:

1. Create green jobs through incentive-based polices, such as the manufacturing and installation of solar panels.

2. Improve public health by removing harmful pollutants from the air and improve water quality.

3. Increase local control by reducing dependence on imported water and energy.

4. Help homebuyers educate themselves about the energy and water usage of a building before purchasing it and without adding significant delays or costs to the home-buying process.

5. Improve public transportation, plant trees, and reduce landfill waste.

6. Save taxpayers money by decreasing municipal water, waste, and energy usage in city-owned buildings (City of San Diego, n.d.).

As part of this plan, in 2016 the mayor's office promoted the "Smart Streetlight Program," not only to the city council but to the general public as well. The most salient feature of the plan was that it could save the taxpayers $2.8 million yearly in energy costs.

The streetlights were labeled "smart" due to their ability to turn on and off, depending on whether or not there were people within a certain distance. The lights also came equipped with multiple Internet of Things sensors, which could generate data such as:

- Number and location of available parking spots on the streets.

- Number of motor vehicles passing by on each street wher lights are installed and also at what speed each vehicle is moving.

- Number of bicycles passing by on each street where lights are installed.

- Number of pedestrians walking on each street where lights are installed.

- Weather measurements in real time of temperature, humidity, and barometric pressure.

Source: City of San Diego

San Diego partnered with General Electric to provide the lights. The Internet of Things sensors in the fixtures continuously uploaded all their data to General Electric servers and to the public via the San Diego website. The data available to the public was referred to as "open data," which was meant to encourage the public, local software developers, and businesses to create new applications, which in turn would further help to improve city services, such as transportation planning and emergency response systems. However, the openness of the data sometimes compromised users' privacy and security, which I will discuss in a later chapter.

Given my technological knowledge, I decided to use the open data to see where it would take me. For several months, I invested many hours developing a geographical information system that pulled data from General Electric's servers via the city of San Diego's interface, which had been made available for public use specifically for researchers to study and analyze.

The idea was that outside input might come up with useful solutions that could help residents.

I faced some initial challenges here and there, mostly in relation to technical support, since the city's technicians with whom I communicated were unable to answer most of my questions relating to data format, authentication, communication, and so on. More than once, they referred me to General Electric's technical support staff. Admittedly, that caused delays, but ultimately I got my questions answered.

I was making good progress with the project until July 8, 2020, when I suddenly could no longer access the data. When I emailed the city to find out why, I learned, ten days later, that all data access had been temporarily suspended because the city was working through an administrative process. However, I was assured that access would be restored by the end of the month.

I thanked the city staff members for their help and asked if they might alert me once the service was restored. By the end of the month, however, I still hadn't received word. When I followed up with another email, I got no reply.

To date, more than a year has passed since I lost access to the data. I might add that I never did receive any updates. Unfortunately, my project failed because I obviously could not analyze data that I did not have.

A few months after I gave up the project, I read an announcement on NBC's website, which stated about San Diego's Smart Streetlights project: "Taxpayers could expect to pay over $2 million more than originally projected, plus expect higher maintenance costs." When I read that, I understood why I no longer had access to San Diego's smart city data.

On February 12, 2020, Erik Caldwell, the Deputy Chief Operating Officer of San Diego's Smart and Sustainable

Communities Program, issued a statement claiming that within the next five years the program's costs would exceed $1.4 million more than had been projected. He added that the city would also need to shell out an additional $1.1 million for "unplanned operational expenses."

	5 Year Forecast - FY21	Current FY 21 Estimate	Increased Cost
Unrealized Energy Savings	$ -	$ 800,000	$ 800,000
Additional Energy Expenses	$ -	$ 700,000	$ 700,000
Operational Expenses	$ 2,273,100	$ 2,191,000	$ (82,100)
Total	$ 2,273,100	$ 3,691,000	$ 1,417,900

Source: City of San Diego

According to Caldwell, "In July 2018, sustainability leadership determined that, due to a lack of oversight and proper due diligence coupled with a limited technological expertise of existing project staff, critical deficiencies had developed with the program." The project suffered from cost overruns, poor supervision, and a lack of properly trained staff, who were unable to manage such large initiatives, even with the aid of current technologies. The city had not taken into account that the new digital infrastructure required a sustainability budget. The program should have been broken down into capital costs (to fund such a large-scale endeavor), as well as allocations set aside for operations and maintenance costs.

Unplanned Operational Expenses

Costs to operate and maintain the smart streetlights and sensors were not planned for when the agreement was executed. Annual costs to operate and maintain the upgraded lights and sensors include:

a. $1.4M/year for API hosting in FY21, increasing to $1.9M/year in FY22 (per node cost from $360-$667/node)
b. $345K/year for data connectivity in FY 21, increasing to $485K/year in FY22 (lump sum data contract with AT&T)
c. $360K/year in Light Grid Hosting in FY21
d. $36k/year in data analyses and application development
e. $50k/year for repair/maintenance costs

To address these issues, we have worked with Current (GE has since sold this business line to Current) to lower the costs of the API hosting from the original quoted amount. We have also worked with AT&T to lower the cost of the data plan needed to transmit the data. For FY20,

Source: City of San Diego

These results were not surprising, considering that the majority of cities today are still governed by politicians who don't have a background in building strategies driven by technological advances. A smart city initiative requires a transformational effort that incorporates both business and technology directed by leaders with vision.

Two professors at Oxford University co-authored an article entitled "Bureaucracy and Democracy in Local Government," in which they examined the current structures of governments at different levels, paying particular attention to municipal governments. "Despite major changes," they wrote, "bureaucratic structures and hierarchies with traditional and modern elements remain a part of the local government organization" (Mossberger et al., 2015).

I believe that cities must lose their rigidity and become more agile in adapting to the new. Today's decision-makers can only make well-informed, rational decisions if they have the right data at the right time.

It is essential that cities embrace Industry 4.0's technological discoveries to provide better living standards for their citizens. Officials need to run their cities like business enterprises, viewing their citizens as shareholders in those enterprises. Of course, that does not mean converting public communities into profit-making private companies. Instead, the officials should emulate business principles, methodologies, and strategies. If they did that, they could more ably train their existing staff and recruit new talent, while creating a culture of innovation and engagement.

Enterprise architectural practices and frameworks like TOGAF were created specifically to avoid the problems that San Diego has undergone. It is reasonable to assert that *all* communities need to adopt the principles of enterprise architecture to conduct their present-day business affairs. Enterprise architecture can enable any communal body to deal more effectively with issues like traffic management, street parking, resource management, urban planning, and so on. When all of these components are integrated with one another on various dimensions, that effectively raises a city's intellect.

Chapter 5
Where Do I Park My Car?

When was the last time you drove up and down a street, looking for a parking spot? And how many of us avoid going to crowded malls and shopping centers because we don't want to deal with the hassle of finding a parking spot? Unfortunately, the solution is not public transportation, because in most cities public transport is still woefully inefficient.

According to a recent study, the never-ending search for parking is costing Americans $73 billion a year. The study took samples from 100,000 locations across 8,700 cities in more than 100 countries to analyze the economic cost of "parking pain" in various nations (Inrix, 2017).

Rank	U.S. City	**Average 2-Hour Parking Cost** (One mile of city center)	**On-Street Search Time** (mins/trip)	**Off-Street Search Time** (mins/trip)	**Annual Search Time** (hours/ driver/year)	**Annual Search Cost Per Driver**	**Annual Search Cost Per City**
1	New York	$33	15	13	107	$2,243	$4.3bn
2	Los Angeles	$14	12	11	85	$1,785	$3.7bn
3	San Francisco	$12	12	11	83	$1,735	$655m
4	Washington D.C.	$18	10	9	65	$1,367	$329m
5	Seattle	$10	9	8	58	$1,205	$490m
6	Chicago	$22	9	8	56	$1,174	$1.3bn
7	Boston	$26	8	8	53	$1,111	$262m
8	Atlanta	$6	8	8	50	$1,043	$251m
9	Dallas	$6	8	8	48	$995	$726m
10	Detroit	$9	6	6	35	$731	$209m
	US	***$4***	***2***	***2***	***17***	***$345***	***$72.7bn***

Source: https://inrix.com

The average U.S. driver spends seventeen hours per year searching for a parking space. That costs $345 per driver per year in wasted time, fuel, and emissions! According to the U.S. Environmental Protection Agency, motor vehicles collectively cause 75 percent of the carbon monoxide pollution throughout the nation. The United States has 30 percent of the world's automobiles, but is responsible for roughly half of the planet's toxic emissions from motorized vehicles. To be fair, the American government is working diligently on this problem.

There is, however, another relevant issue that could alleviate the pollution quandary: parking. Solving the parking problem could significantly reduce toxic waste. Unfortunately, it is a concern that has not received the attention it deserves.

Cities around the world struggle with the issue of insufficient parking spaces. The continuously rising populations have led to an ever-increasing number of vehicles on the roads. In addition, more people than ever can now afford a vehicle and have the freedom to drive to and from their destinations. Cars have become essential to our daily lives. For that reason, the number of registered vehicles now exceeds that of the population. This disproportionate number of motor vehicles on our roads has intensified the search for parking spaces. Parking and all its attendant problems have now become a prime issue in daily urban life.

I usually arrive on campus at least ten to fifteen minutes early to ensure finding a parking spot before my class. It is a good day when I find parking right next to the engineering building, where my class is held. On those days, it only takes me two or three minutes to walk to class. Unfortunately, those days are rare.

One day, when I had class at 7:00 P.M., I arrived ten minutes early to be there on time. Unfortunately, the parking lot was full that day. I kept circling around, hoping that someone would leave. Finally, I had to leave the first parking lot for another one, which was further away.

By now, my search had made me ten minutes late. Fortunately, I spotted a woman getting into her car and took her spot. I now had spent twenty minutes looking for a parking space, and it took me another four minutes to walk to my class.

The moment I stepped into the classroom, some of the students were preparing to leave, thinking I was a no-show. As I began the class, I did my best to hide my frustration. After telling the approximately seventy students why I was late, I asked them what was the longest time any of them had taken to search for a parking space? Their overall average time was twenty-five to thirty minutes. The worst-case scenarios, up to an hour of searching, occurred at peak business times.

As I was teaching that class, the issue of parking remained on my mind for quite a while. I had to wonder, What is the city doing about this?

On the Long Beach's city website, its new parking app proclaimed, "As you probably know by now, I'm the free app that shows you—in real time—all the open, available parking spots in Long Beach." The city's "This-will-solve-your-parking-problems" campaign was for its new app called "EZparkLB." So, I downloaded the app to test it out, eager to see if it could help solve my problems at the school or anywhere else.

The app was designed to bring "visibility" to some of the parking lots in the city. It operates by locating either a parking structure or an open lot near your location. If the data is available, the app can also tell you the fee required. One drawback, however, is that it completely ignores parking meters and street parking.

The information can be accessed by searching on Google for "parking lot near me." EZparkLB was designed to work only within the borders of Long Beach. However, if you go to Los Angeles, there is another app available there, and so on for New York, Washington, San Francisco, and so on. Before you know it, your phone will be filled with hundreds of parking apps, based on how many cities you travel to!

The most common problem that drivers have with parking lots is finding out whether or not there is a parking space available in them. I circled around in my school's parking lots for fifteen to twenty minutes, not because I didn't know where the parking lots were, but because I was looking for an available spot.

Parking visibility has been a problem in all U.S. cities for some time, but as of now, there is no solution. Providing a universal solution requires a nationwide strategy that engages multiple cities, so that they can collaborate. One of the most contentious problems we are facing today is that our cities are still thinking and working in silos as individual entities. Operating within that modality will continue to hinder any community from successfully transforming itself for Industry 4.0.

These outmoded municipal systems all lack the basic principles of enterprise architecture that I discussed in Chapter 2. As I have noted, cities should emulate a business model, in which integration would span across every business unit within an enterprise. To place the issue in another context, if we think of the whole country as an enterprise from a systemic standpoint, then the nation's cities are organizational units within that enterprise, all sharing common goals and objectives. In this context, one of the principal goals of such an enterprise would be to solve the parking problem, which would in turn reduce pollution and result in a better living experience for the shareholders—who, in this case, are the citizens of the various communities.

Today, however, each city views its parking issues only from the perspective of its municipal districts. But people today are constantly moving around, especially during the prime hours of the day. For instance, I live in Long Beach, but if I worked in Los Angeles, on my commute back home I might stop in Torrance for a doctor's appointment and then go to Seal Beach for dinner.

If each city only solved the parking problem for itself, they would each have to create a mobile app for their own territory. So, someone living within the Los Angeles Basin would have to download multiple apps for Long Beach, Los Angeles, Torrance, Seal Beach, and possibly more. Ostensibly, a person traveling on an interstate highway for work or vacation would need a multitude of city apps to find parking spots in each locality. With forty-eight contiguous states and hundreds of cities and towns, that would amount to dozens or hundreds of apps on any one phone! Can you think of anyone who would like that?

Community Engagement

Some parking data within just one city is broken down into various districts. For instance, in Long Beach, some parking structures that are managed by the city have payment machines provided by a company called "LAZ." LAZ provides Long Beach with data about parking availability, but the city does nothing with that data. LAZ created a parking app, which provides visibility for any nearby vacant parking spot, but only within lots that utilize its equipment. Some lots have equipment from other vendors. In any case, neither the city's app nor LAZ has yet to provide information about the availability of parking meters on the streets. It is convenient that the LAZ app allows you to pay for parking with your phone, but the app still does not solve the overall parking problem.

The list of apps goes on and on for each city or company that holds a small piece of the overall puzzle. Solving such puzzles is at the core of Industry 4.0, and one of the reasons why any city that uses such a solution would be deemed "smart."

As a concerned citizen who wanted to help solve the problem, I decided to do the necessary due diligence to dig deeper into the subject. Fortunately, I had already developed a system that was based on real-time Internet of Things sensor data for parking availability—the same system that I had used to access data from San Diego's website. Unfortunately, the data I needed access to was no longer available. Also, the system I had formerly used applied only to the city of San Diego. Nevertheless, I forged ahead, using the strategy I had used to help the city of Long Beach adopt and implement changes with its parking issues.

I wondered what other cities I could help in the same way, probably starting with Los Angeles itself. That thought inspired me to do what I could to at least solve some of the daily problems that we all face, including parking, traffic, pollution, and so on. An enormous number of solutions are within reach these days if we use the technological wizardry that Industry 4.0 has given us.

Sunflower Seeds

I set up my first business when I was only 10 years old. It began as a small operation, selling sunflower seeds to my nine brothers and sisters. When I first moved to the United States, a classmate asked me, "Yo! Malik! How many siblings do you have?"

MALIK: I used to have nine, but I lost a brother, and now we're nine.

CLASSMATE: Hold on one sec. Did you say nine?

MALIK: Yes.

CLASSMATE: All from the same mom and dad?

MALIK: Yeah.

CLASSMATE: Oh, my god! That's almost a basketball team! How can you remember all their names?

MALIK: Because they're my brothers and sisters!

After my classmate's stunned reaction, I answered that question with "three siblings," since I didn't want to go through the whole rigmarole again. Anyway, my family was overseas, so no one knew the difference.

Rearing a family of ten kids is financially challenging enough, but raising them with a single income is a major hurdle. When I was a kid, large families were common. My dad, who was a truck driver, often took trips that lasted for days at a time as he moved goods back and forth to neighboring countries. My mother stayed at home, took care of the children, and made sure we all went to school and did our homework on time.

One of our greatest sources of pleasure was that she used to give us daily pocket money. We each got a small allowance, except for those of us who had already graduated from college and gotten a job. The tiny allowance was enough to buy candy or a small bag of sunflower seeds. Eating those seeds as we watched cartoons on TV was one of our favorite treats. At night, when my parents watched TV, they also liked to munch on the seeds.

One of our neighbors, who lived a few blocks away, had

converted a small room of her house into a little market, where she sold sunflower seeds and candies. One day, I walked into the shop and watched her roasting the raw seeds. After that, she put them into bags for sale. That set off a light bulb in my head. After seeing how easy it was to do, I decided to roast sunflower seeds myself and sell them to members of my family. I figured they might as well buy them from me, instead of walking to the candy shop.

I began saving my allowance until I had enough money to buy half a kilogram of raw seeds. While I was saving up, I often went to play with the children of the owner of the candy shop, keeping an eye on how she prepared the seeds for sale. It wasn't difficult. She started by rinsing the seeds in water to remove any dust, and then she sprinkled a lot of salt on them. She then left the salted seeds out in the sun on an open tray for a whole day. After they dried out, she roasted them with low heat in a gas stove, turning them over from time to time until they were brown. When she could smell their toasty aroma wafting in the air, she knew they were ready.

I began my business one day by asking my mom if I could go with her to the market.

"Why?" she asked.

"Because I want to buy raw sunflower seeds," I replied.

Then I told her my plan to roast and selling them to family member.

"Instead of us going out to buy them from the shop," I said, "everyone can buy them from me!"

I was surprised when she didn't object.

"You'll buy them with your own money?" she asked.

"Yes!" I said proudly.

The plan went as I expected, and soon I was selling sunflower seeds to my family. Not long after that, I began to make a profit. After a few months, I announced to the kids in the neighborhood that I had sunflower seeds for sale, and before I knew it, customers were coming in and out of the house all the time.

I also started going to the market on my own because I needed supplies every two days, instead of every two weeks. On one trip, I added candy to my purchase, so my business expanded to selling roasted sunflower seeds and candy.

However, my expansion came with a price. Since kids were knocking on the door all the time to buy seeds or candy, my dad got annoyed one day with the constant knocking, and decided that enough was enough.

"This needs to stop!" he cried. "You need to focus on your schoolwork. I don't want to see or hear these kids at the door anymore!"

That was the end of my first business enterprise. However, less than a month later, I found other ways to make money, which I needed to cover my school expenses. I went from selling seeds and candy to selling ice cream, then car oil, and then lemons.

Right after I moved to the United States in my early twenties, I worked for a company whose owner was a very greedy man, which gave me the impression that all business owners were like him. In fact, I actually came to hate the idea of owning a business. Unlike the little business enterprise I had had, real business enterprises seemed to be on a whole different level. For a while, I even believed that owning a business might turn me into an avaricious character like the owner of the business I had worked for.

Of course, there is a difference between owning a business and operating it. When you both own and operate an enterprise, that can create a balance between making a honest profit and just being greedy. Believing that I could strike a balance between the two, I launched my own business.

Starting Up

After spending days trying to come up with a business name, I began to wonder if I would ever find one that hadn't already been taken. Every domain name I came up with already belonged to someone else. When one buys a domain name from Google, GoDaddy, or whomever, the first step is to type in the website name you have in mind. A search then tells you if the name is available or not. If the search results indicate that a name is taken, names of broker services usually pop up with lines like:

> "Domain Broker Service"
> "We still might be able to get it for you"
> "We'll contact the owner and negotiate on your behalf"

I discovered that some enterprising individuals and companies had made a business out of offering people website names. With the help of computer algorithms, they were able to generate any number of potential names with three, four, five, or six letters, since most people like to have short names for their website. For this service, one might be charged an annual fee between $30 and $50.

A good thing in my favor was that my last name, Luti, is not a typical Middle Eastern name, most of which have at least eight to ten letters and are difficult to pronounce in English, so people often misspell them.

Since Luti.com was already taken, I checked with Domain Broker to see how much it would cost to buy it from the owner.

The broker responded that the owner was asking $50,000, but was open to a counter offer. I countered with a bid of $500, which was immediately declined, so I eventually chose a different name—for which I paid nothing. This name game is a new industry that didn't even exist twenty years ago.

Of course, today the process of opening a business no longer requires one to register with a government agency. Companies like incfile.com have automated the whole process with a few clicks. You only need to visit their website and choose the city and state where your business will operate. For a modest fee, they will process the paperwork on your behalf. They can also send any papers you may require to your home—all without the need to go anywhere.

Getting to Know My City Representative

Since moving from Seattle to Long Beach, California, in 2015, I have never had a reason to go into city hall except to pay for an occasional parking ticket—usually received only moments after the meter had expired.

In fact, I never felt the need to know who my city councilor was or what his or her responsibilities were. It just never occurred to me to question what the city was doing to solve day-to-day problems like traffic management, city planning, and so on. I simply trusted that things were going smoothly, and that there was no need to get involved. However, I still grumbled whenever I drove to the downtown area and ended up circling around and around for a parking space. I also wasn't a happy camper when I had to slog through the colossal traffic from Los Angeles to Long Beach every time I returned from a doctor's appointment.

Somewhere along the line, I learned that Suzie Price was the councilwoman for my district in Long Beach. I was delighted when I learned that she was working on getting approval for smart parking meters in the Belmont Shore area, where I live. That was good news because the old meters could only be paid with coins, whereas these new ones would accept credit cards. That was wonderful, but was it enough? Although the new meters would be convenient for payment, they would not solve the bigger problem of parking visibility and availability.

I contacted Suzie's office to ask if I could have a meeting with her to discuss a potential solution I wanted to suggest regarding Long Beach's parking issue. After two weeks, I got a reply from her office, scheduling a meeting in ten days.

At the meeting, I thanked Ms. Price for her work in the community, and then brought up the topic of the parking meter replacement project, proposing an amendment to it that divided the parking problem into two different parts.

First, there is the problem of parking visibility leading to a blind search strategy. Many drivers in the city, I said, including myself, prefer metered parking spaces on the street to parking lots because metered spaces are less expensive. However, that often means that we have to drive around in circles looking for a space. That familiar pattern is what I call "searching blindly."

I pointed out how the city's current system did not provide information about meter locations. Then I showed her four pictures that I had taken at two different locations. The first picture showed an empty parking space with a meter. The second picture was a screenshot of the city's parking app, which clearly showed that it wasn't even detecting the very spot I was standing next to! The third picture showed a full parking lot on the ocean. The fourth picture was a screenshot of the city's parking app, which did not detect the lot at all.

2nd and Granded

Meter parking in not available

EZparkLB

Long Beach Parking Lot

This Parking lot was not detected

EZparkLB

Second, there is the problem of parking availability leading to multiple cars chasing a single space. I noted that people were wasting a lot of time driving around looking for available spots, which was not only adding to the daily stress of life, but also polluting the air.

Then I handed her two more photos, both by the ocean, which showed a scene I see almost every weekday, which gets even worse on weekends. In the picture, people are entering a parking lot, looking for an available spot. They have to go up and down, hoping to see someone leaving, but most of the time, they just give up. When they leave the lot, they pass a long line of drivers just like them, who are waiting to get in and have to go through the same rigmarole that they have just gone through.

I then brought up the city's new initiative to make Long Beach a "smart city" by using some of the latest technology, suggesting that that initiative could be specifically applied to the parking problems in the city. "After all," I said, "isn't the goal of the initiative to improve the overall living standards of the city's residents?"

After acknowledging the problems, Ms. Price suggested that I contact a Mr. Logan (not his real name), the director of the Long Beach Public Works Department, who was responsible for such issues. Her chief of staff then sent out an email the next day, introducing me to Mr. Logan and asking him to schedule a meeting with me to discuss my ideas. The email was sent on July 30th, but after a week, I hadn't received a response. When I sent a follow-up email, there was no response to that either.

In the meantime, I decided to work on my sales and negotiation skills, hoping to convince the city to implement my initiative. I read any book I could get access to that related to the topic. Then I came across Never Split the Difference by Chris Voss, a former FBI negotiator, who describes clever tactics that one can use to get any negotiation moving forward. They were the kinds of suggestions that one can use in both one's professional and personal life.

One of Voss's techniques makes people resume a negotiation after they have dropped the ball. The tactic was to use a simple point-blank question: "Have you given up on this project?" An email with a subject title like that can reactivate a dialogue that has gone dormant. Voss lists other techniques that can compel opposing parties to resume their participation in the dialogue, often within minutes!

In my case, Mr. Logan and I had never spoken to each other, so I was already at a disadvantage. After waiting for a month and a half, I decided to try Mr. Voss's idea. I emailed Mr. Logan with the question, "Hello, Mr. Logan, I am wondering if the city has given up on making Long Beach a smart city?"

Boom! I got an immediate response! I sent my email at 10:32 A.M., and I received a response by 10:37! The response also came with a friendly "good day" and a suggestion for a time to meet.

I had anxiously waited for over a month and a half with no response. Yet, when I followed Mr. Voss's advice, I got a reply within five minutes! It was like magic. Thank you, Mr. Voss!

On the day of our meeting, Mr. Logan showed up slightly late. He arrived accompanied by a Ms. Warden, who was the project officer responsible for the city's parking operations. Mr. Logan mentioned that Ms. Warden held a bachelor's degree in criminology as well as an MBA. Then he explained that he didn't have much knowledge in this area, and that Ms. Warden made the final decisions in such matters.

When it was my turn to speak, I began by identifying the problem I came to talk about, and then I identified the city's current way of handling it, pointing out what I believed were obvious drawbacks and why they were problematic. At the end of my presentation, I offered my solution.

What I said seemed to have rung a bell for both of them, after which I answered their questions. At the end of our meeting, Ms. Warden remarked that she would need time to digest all the information I had given her, and then she would get back to me soon and let me know what would happen next.

Lack of Vision

A week later, I received an email from Ms. Warden, in which she again acknowledged the problem and the need for a solution. However, she then made a disappointing comment, noting that the city would continue to work with its current vendor, INRIX, to solve any issues related to our discussion. When I heard that, my heart sank.

INRIX, which is based in Seattle, provides location data and analytics to several major cities about parking information, traffic patterns, and so on. INRIX also works with luxury car manufacturers, such as BMW. With its data, INRIX assists the car manufacturers to create software that helps car buyers to find parking spaces. For instance, to attract more car buyers, INRIX and BMW signed a deal in 2018 to provide BMW owners with software that contained a machine-learning algorithm that locates available parking spaces, based on information from BMW owners and cities like Long Beach. The software also helps to cut down on cost, driving time, fuel emissions, and so on.

This poses the question: What about those of us who can't afford a BMW? Must we own a luxury car to get ready access to parking information?

The question anyone should ask is: What is my city's strategy for solving the parking problem? Is Long Beach working with the state to implement a solution that can work across the whole of California? What about other states? Is there a nationwide strategy to solve this issue? Once again, searching for parking spaces costs Americans $73 billion a year, and the average driver spends seventeen hours per year searching for parking.

Recently, the city of Long Beach posted the image below on its Instagram account, comparing today's current parking problems with the horse and buggy problems in 1905! The photo was accompanied by the statement: "If you think parking is bad now, check out the horse and buggy situation that happened in Long Beach in 1905."

10:13

LONGBEACHLOCALNEWS

Posts

longbeachlocalnews ...

1,134 likes

longbeachlocalnews If you think parking is bad now, check out the horse and buggy situation that happened in Long Beach in 1905. OP : @fill_in_the_blank_jones #longbeachlocalnews #lbln #longbeach #throwbackthursday #tbt

View all 33 comments

February 4

Twenty years ago, discussions about a nationwide strategy to solve parking problems would have been impossible, because there were not yet any technologies to enable such strategies. Until now, the federal government has relied on cities to implement any such strategies, but that approach has not worked. Letting communities develop their autonomous solutions in no way contributes to any sort of nationwide strategy.

Bear in mind that resolving the parking problem would also mean that citizens would, for the most part, no longer get parking tickets. With an app, they would not only know exactly where to park, but they would also receive a notification on their phone about when their time was expiring. Furthermore, they would notified about such issues as street sweeping times, during which parking is not allowed.

On the other hand, a massive reduction of parking violations would cut into a prime source of income for most cities. According to a recent study, annual parking revenues reach $1.4 billion (Keiser, Keilah, 2019). Do you think any city is ready for such losses?

Earlier, I went into detail about the digital transformation and how many companies are currently remodeling their infrastructures and rehauling their business models to cope with Industry 4.0. I gave examples of other countries that have launched national strategies to undergo such efforts by transforming the way they dispense services to their publics.

As of now, there is no clear strategy in the United States to prepare either the government or the population at large for such a transformation. Of course, I am not just referring to issues like the parking space problem. Today, issues beleaguer us at every turn that will require nothing less than a nationwide strategy to be resolved.

On the front burners are the never-ending predicaments of traffic, energy, election fraud, water quality, air quality, and health care. The list of civic woes goes on ad infinitum. However, in matters of national security, within agencies like the CIA, the FBI, or the Department of Defense, technological advances abound. I previously mentioned the U.S. Air Force as an example. That one branch of the military has activated a massive effort to redesign its entire infrastructure. It is also undertaking ambitious retraining of its personnel from top to bottom. From all appearances, these agencies seem to be on top of the digital transformation.

I fervently believe that the average citizen, who deserves to have better service, will only ever receive it if and when we undergo a nationwide strategy that enables cities to transform their staffs and infrastructures to accommodate Industry 4.0. Although Industry 4.0 is inextricably associated with technology, it is also important that we nurture the right culture for such technology, so that we will be able to gain the fullest advantages from its vast potential.

In the next chapter, I will provide another example of how technology has the potential to make our lives easier, and how the lack of leadership, knowledge, and bureaucracy is hindering that.

Chapter 6
A Peaceful Stroll

"Yo! On your right!"

"Don't ride so fast!"

"You're not walking in your lane!"

"Please show a little more respect. Stay in your lane!"

"Stop being so f—king disrespectful!"

"One day you're gonna get run over if you keep crossing your line!"

I hear shouts like this all the time while riding my bike on the beach trails. People cross from pedestrian to bike lanes back and forth, with everyone carping at everyone else. Sometimes people even get hit by a bicycle, and an ambulance has to be called. These things happen almost every day. But who is really at fault? Is it the pedestrians who are walking in the bike lane, or the bicyclists who are riding in the pedestrian lane? Or perhaps it is a bicyclist who is riding in the wrong direction.

I took the pictures below while I was on my daily bike ride around Long Beach, specifically in Shore Line Village Park.

It's sad to see people getting hurt. It's also sad to constantly hear heated words flying back and forth between people. Ironically, these disputes all take place over issues that none of them have any control over.

The concrete pathways in Shore Line Village Park have three lanes: two for bicycles going in opposite directions and one for pedestrians. The cause of all the tension is that the width of the lanes, which were designed years ago, is not wide enough for both bikers and pedestrians. At the time that the concrete was poured, the future growth of the population was not taken into consideration.

The figure below illustrates how Long Beach's population increased from 361,500 in 1982, the year the park was opened, to 456,154 in 2021. The steady increase in population, along with the growing number of daily visitors, was not factored in when the lanes were poured.

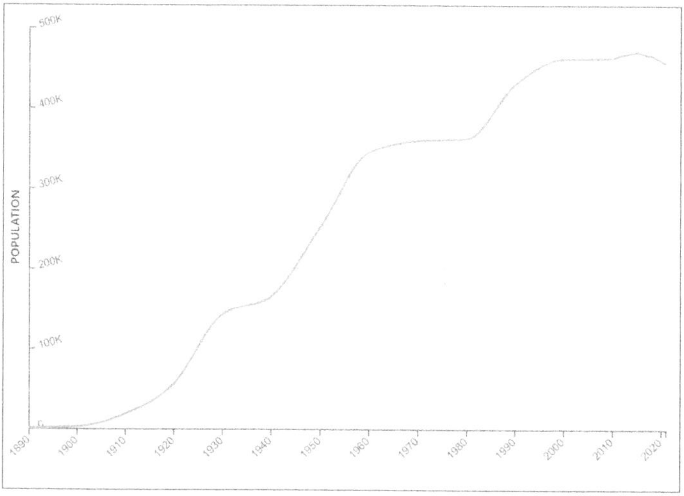

Source: https://www.worldpopulationreview.com/

I was curious to know if the city was even aware that these lanes were not suited to the number of bike riders and pedestrians using them, so I decided to call the city officials to find out two things. First, I wanted to learn if they knew that their streets were not big enough for the current population, and if they were aware of the injuries and growing friction that the inadequate lanes were causing. Second, I wanted to know how the city determined the size of the lanes and how temporary the determination of future population growth was.

I found the city's phone number by looking it up on Google, and began the slog that one has to go through to find out things like this. When I made the call, the first thirty minutes were spent just getting transferred from one department to another. I was just about to hang up when a customer support person answered and gave me an email address for someone in the Public Works Department, who would be the best possible contact for getting my questions answered.

Next, I exchanged some emails with a man I'll refer to here

as Mr. Cassidy. At one point, I actually had a phone call with him to discuss the issue. But I came away realizing that the city didn't know anything about these matters or just how bad they had become.

From Mr. Cassidy, I learned that there is an equation that engineers use to calculate the varying widths of public walkways or bike lanes. That equation is based on the number of pedestrians observed over a given period. Mr. Cassidy explained that the city sends someone out once a year to update that number. To my surprise, he said it was done entirely by human observation. Then he went on to say that he was aware that such important statistics could be determined much more accurately by utilizing new technologies, but he had failed to convince the city managers to invest in such technologies.

"You seem knowledgeable about the topic," he said. "Do you work in this field?"

I told him that I teach at California State University in Long Beach and was conducting research about Smart City initiatives. I added that I was building a platform that would help the city to provide better services for its residents.

Intrigued, he asked, "Would you be interested in giving us a demo for your platform?"

"Of course!" I replied.

"Then I'll check with my manager," he said, "and we'll send you the date and time for the meeting."

After a month had passed, I still hadn't heard back from him.

So, I sent a follow-up email, asking about the status of our meeting.

He replied, "I have a new supervisor, who just started in our division last week. Let me forward this email thread and ask her to make time for your demo. I'm going to also invite the city's data scientist to attend. Sound good to you?"

Delighted, I said, "Yes, of course!"

Unfortunately, months later, I still hadn't heard back from him or his manager.

I'm sure the problem was not with Mr. Cassidy, since he sounded motivated and open-minded. Most likely, his manager was resistant to hearing or implementing new ideas. Meanwhile, the tragedy of people getting injured and the continuous bickering continues to worsen.

Streets That Can Sense

There is no way for one person, solely using his or her sense of sight, to arrive at an accurate count of pedestrians passing by on a street. The number will vary according to the time of day, the day of the week, holidays, and so on. Today, in Industry 4.0, there is a wide range of physical objects from motor-driven vehicles to surveillance cameras to streetlights that are endowed with the ability to sense and interact with humans. It's up to us to listen, understand, and respond to them in a way that can make our lives easier.

In the case of the width of the pathway lanes, if the city installed Internet of Things sensors in the streets, it would have access to a continuously updated count of the pedestrians passing by each minute, hour, and day. Those sensors would relay real-time data to the city's servers. The data could also help engineers to forecast many engineering challenges ahead of time.

Adopting artificial intelligence tools now will enable us to shape our future responsibly. In these early days of the Fourth Industrial Revolution, we must get this right while there is still time. So why is that not happening?

The Challenges of Cities in the Digital Information Age

I mentioned earlier how the whole process of digital transformation has upended entire industries. Businesses have been forced to take on completely new identities to keep up with the rapid-fire changes being brought on by Industry 4.0. If enterprises don't stay up to speed, they soon find themselves out of business.

As a matter of survival, we should all be resoundingly aware of this current industrial revolution, asking ourselves how and why we need to adapt to it.

Corporations such as Walmart, General Electric, Airbus, and Lockheed Martin, as well as governmental agencies such as the Defense Department, the CIA, and the FBI, are aware of the urgency of this matter. Therefore, many of these entities have upgraded their infrastructures, invested in research, and begun to train their workforces to keep up with the demands for the new skill sets needed. Their predecessors, such as Blockbuster, Sears, Westinghouse, Kodak, and Borders, all ended up like lumbering behemoths fading into extinction because they had lost their agility and adaptability to change with the times.

The constant development of new technologies and the scale of challenges are mounting daily. We should be concerned about and involved in how our cities and municipalities are keeping up with all of this. Scientific and technological advances have disrupted our day-to-day lives in many ways, which should be of profound personal concern to everyone. As residents of a community, we are not unlike shareholders in an enterprise. Like shareholders, we need to practice our right to participate in the supervision of our cities. We need to ensure that our cities' projects and initiatives are built and developed according to our needs.

In Chapter 4, I highlighted some of the challenges that cities and municipalities face. Those challenges include leadership, agility, and citizen engagement.

Leadership

A great deal has been written about the importance of leadership. Suffice it to say that leadership plays an important role in the development of any organization. I think that we can all agree that good leaders can inspire and guide the members of their organization effectively. With competent leadership, a well-run organization achieves smoother productivity and can more ably maximize its efficiency.

The best model of leadership is leading by example. At the same time, it is one of the most challenging models. John Wooden, the head coach of UCLA's basketball team, said, "A leader's most powerful ally is his or her own example. Leaders don't just talk about doing something, they do it" (Geier, 2019).

In a technological revolution such as Industry 4.0, the leaders of any municipality need to be digitally literate, not just politically savvy. Digital leadership requires using digital assets and understanding technology to achieve goals. It is not only about lobbying for new technologies, but also about completely overhauling organizational structures, governance, and work processes, and laying the proper groundwork for the right culture. Only then can governmental leaders align their functions successfully with the digital transformation that is under way. Today's government leaders must be informed and ready with forceful arguments on behalf of digital transformation. Ideally, they must be continually expanding their skill sets in digital literacy to match the competency of the private sector. People today expect public services to be as personalized and responsive as the services they get from any commercial enterprise.

There is already a disconnect between local government leaders and the residents of the communities they serve. To connect with the public, those leaders should have the mindset of a "citizen-first" culture when designing city policies and delivering services. Given what I know about current technology, I am convinced that it can play a huge role in bringing governments closer to people and their actual needs. But that technological transformation will only happen if it is implemented in the right way. Are our community leaders ready to take on these exciting possibilities?

Leadership Makes "Smart City" a Buzzword. In October 2019, the mayor of Long Beach, Robert Garcia, and the City Council directed the city staff to develop a Smart City Initiative. One of the first steps in that initiative was to place a statement on the city's website: "We think technology and data have the power to improve the lives of all Long Beach's residents and that real change can only happen if we collaborate as one city to solve problems and test out solutions." The statement went on to identify four main principles in the implementation of the initiative:

1. *Design for Equity:* Reduce historic inequities by ensuring technological advancements are accessible to all, and further improve the quality of life for those communities that have been underserved.

2. *Earn Public Trust:* Build public confidence with secure data privacy, data transparency, and community engagement.

3. *Cultivate Local Expertise:* Promote place-based growth by supporting Long Beach entrepreneurs and businesses. Improve workforce job-readiness, and build partnership networks.

4. *Build Civic Resilience:* Improve response time to the most pressing civic challenges using data-informed decision-making. (City of Long Beach, n.d.)

Principles of Enterprise Architecture. In the practice of enterprise architecture, the principles are considered the guides that influence the development of any strategy. The Open Group organization defined those principles as: "general rules and guidelines, intended to be enduring and seldom amended, that inform and support how the organization sets about fulfilling its mission" (The Open Group, n.d.). Although the principles are intended to be long-term and not prone to frequent amendments, an exception is made if there is a valid reason. In such a case, the development team of the strategy and the shareholders both need to agree to the change. The revision is then republished, placing everyone, so to speak, on the same page.

Deviating from Principles. I was excited when I heard about Long Beach's Smart City initiative, especially since I had made progress with my research and was prepared to submit a patent for a smart city system, so I decided to contact the city to explain my interest in supporting its initiative. As a resident, I would fulfill one of the founding principles of Long Beach's Smart City initiative—namely, "encouraging participation from local experts and entrepreneurs, thus improving public trust through community engagement."

After placing a phone call to the city, I was transferred

from one department to another. To make a long story short, I eventually secured a Zoom meeting with a Mr. Longfellow (not his real name), the Smart City program manager. Once we were face-to-face on Zoom, I started the conversation off by thanking him and the city staff for launching the initiative. Then I began my demonstration by proposing potential solutions for some of the critical issues in the city, such as smart parking, traffic monitoring, and automated incident responses. I gave credit where credit was due by pointing out that these solutions had been created by current technology that is integral to Industry 4.0, adding that I had submitted a patent for my innovation.

Mr. Longfellow, who had no technical background, responded, "This is not how we do business. We work in silos, and your solutions require integration across multiple business platforms."

I diplomatically replied that I was aware of that challenge, and then I broached the sensitive topic that for digital transformation to work at all, it required integrating and sharing multiple data sources from different departments. With that in place, I noted, the city would be empowered to respond to the public's needs.

Then I mentioned that it was possible to run a free pilot program to test some of the solutions. Such a test could evaluate the proposition and demonstrate to what degree it was aligned with the city's strategy.

With a note of misgiving in his voice, Mr. Longfellow said, "Hmmm. Maybe you should speak to Dr. Goldsmith [not her real name], the chair of our Technology and Innovation Commission. If she approves your plan, then maybe we can run a pilot program."

I believed that I had made some progress, and was certain that the chair of the Technology and Innovation Commission would surely appreciate any solution that could add value to people's lives. I was convinced that she would be receptive to my ideas. How naïve I was!

I sent out an email to Dr. Goldsmith, introducing myself, and explaining what I had been up to, mentioning the meeting I had had with Mr. Longfellow. Then I requested a Zoom meeting with her to further discuss my proposals. After several days, I still hadn't heard from her. While waiting for a response, I thought of all the people stuck in traffic each day and all the cars lined up at the beach, waiting for someone to vacate a parking space. I also thought about the air pollution and the growing friction between people in the bike and pedestrian lanes. All I could focus on was how much time had already passed without anything being done about the problems. I sensed that such issues were not being addressed because of sluggish bureaucrats and city fathers who were detached from today's reality. Not wanting to wait any longer, I decided to use Chris Voss's magic trick again to expedite a response.

"Hello, Dr. Goldsmith," I wrote in an email, "I am following up on my previous email, and am wondering if the city has given up on the smart city initiative."

Once again, the magic wording worked. I sent the email at 10:47 P.M., and I received a response at 4:21 A.M. with a proposed date and time to meet her: a 30-minute meeting on that Friday at 1:00 P.M.

Dr. Goldsmith started our meeting with, "I have an appointment, so I need to leave ten minutes early."

I responded, "That's fine, I'll make it quick."

Then I came straight to the point, identifying some issues that people were facing and outlining my solutions to the problems.

"Hmmm," she said. "It seems interesting, but I don't have

much experience in technology."

"Forgive me," I replied. "I just want to make sure that you are the chair of the Technology and Innovation Commission. Perhaps I'm reaching out to the wrong person."

She replied, "Unfortunately, I'm not in a position to influence the city of Long Beach to adopt your platform!"

Her words perplexed me, for it seemed obvious that her negative response was not caused by my solution not having value. If that were the case, she would have said so. But I had trouble believing that someone with such a prestigious title could not influence some changes that could very likely make a difference in people's lives. Weren't the city's principles founded on earning public trust and cultivating local expertise?

I began to reconsider the city's manifesto, which now seemed to be made up of high-sounding but empty words. How could the city's officials hope to engender a belief in these principles if their willingness to participate in possible solutions was negated at every turn?

As the disparity between what was said and what was done widened, the public's trust in local government would continue to weaken. Trust is something that takes a long time to cultivate, and dismissing interested and involved citizens is not the way to go about it.

Incidentally, I later learned that Dr. Goldsmith, who made it clear that she did not know much about technology, has a Ph.D. in mass media and communication. She also teaches course in Journalism and Public Relations at a local university. Having spent twenty years of my career, not only in academia but in small, medium, and large organizations, I found it curious that someone who headed a city's department of technology and innovation had no background in technology. Considering this one example, is it any wonder why so many governmental agencies are routinely hacked these days?

Keep in mind that the response I got is representative not only of Long Beach but also of communities everywhere today. Incompetent leadership is one of the main impediments to cities' progress and growth today.

In 1939, the New York World's Fair had as its theme "The World of Tomorrow." Its opening slogan was "Dawn of a New Day." With the advent of all that was new and exciting in the world of technology, the public was shown what was possible at the time. Narrated films predicted that cities would become models of efficiency, maintaining continuous friendly communication with their residents.

The fabulous vision of the future was supposed to be realized by the 1960s. What was possible from the perspective of the late 1930s can now be seen as naïve and simplistic. Today, we don't live in cities that lack traffic accidents or accommodate all the needs of their citizens. But we do have technology that is far beyond the imagination of the 1930s. Ironically, although mighty engines of power are at last at our disposal, most communities have been slow to integrate them into city policies. By contrast, the private sector has eagerly adapted anything and everything that can advance its causes.

Agility

The burgeoning populations of our cities have given the term *urban crisis* a new meaning. Traditional methods of doing business are failing to keep up with the insatiable demand for new products and services. Communities of all sizes need to become more adept at change, reformulating the way they think, operate, and lead to achieve better mission outcomes.

There is an urgent need to apply today's specialized

technologies to vital issues related to society, ecology, and morphology. But utilizing technology is not the only way to improve people's lives. A survey published in 2018 by McKinsey, the global consulting firm, revealed that the shortcomings in any organizational culture are among the main impediments to its successfully adapting to the digital age (McKinsey, 2020). The survey identified three cultural deficiencies:

- Functional and departmental silos
- Fear of risk-taking
- Difficulty responding to and acting on a single view of a customer

These same deficits hinder cities from effectively moving forward. All communities must now shift from a linear to an interactive approach to develop and deliver programs for today's problems. That can only be done by becoming more adaptive, creative, and resilient, responding quickly to the turbulent environments and technological advances that are part of Industry 4.0.

Citizens' Engagement

For cities to be smart and move forward into the new digital age, there must be civic engagement. Having an open line of communication with the public will serve cities as well as their citizens. Without that open communication, the gap between community governments and their citizens will widen even further. Computers and the Internet are becoming more and more crucial in forging a link between the two. Citizens will need to become more proactive by familiarizing themselves with upcoming civic projects.

I need to stress once again how important it is that the residents of communities think of themselves as shareholders. They must see that their engagement is a precondition for sustainable urban planning. Until I started my research, I didn't even know who my district representative was, as I'm sure most people don't. To put the matter succinctly, that is not okay. At any given point, multiple issues are directly impacting our lives in terms of pollution, health care, information security, and so on. We must all become a part of these discussions.

City governments need to encourage citizen engagement because it provides huge benefits for everyone, for at least three reasons:

- Transparency in public decision-making increases trust in local governments.

- Inviting the public's ideas, concerns, and local expertise leads to a collective intelligence between governments and citizens.

- Cities' solutions of residents' problems bridge the communication gaps between citizens and officials.

After spending almost two years working with my local city, my observation is that city officials have not made it easier for citizens to engage. It took me up to three months just to demonstrate to city officials a solution that could have hugely benefitted the community.

I once sent out three emails to the mayor of Long Beach, asking for a virtual meeting by using the "contact the mayor" button on the city's website. A year passed by without any response. As shareholders of the city, citizens need to encourage their city officials to take ownership of sustainable mobility ideas, policies, projects, and so on. That could provide a great opportunity for city administrators to utilize local expertise. Acquiring feedback from interested and concerned citizens would greatly improve the quality of cities' services.

Part 3
Smart Citizens

From here on, I will detail how we can cope with the Fourth Industrial Revolution. Whether we are aware of it or not, we are all very much a part of this social and economic phenomenon, which impacts our lives every day by compromising our privacy, automating our jobs, and radically affecting our educational system. So much of our lives is being changed that it is now more important than ever for us to maintain public policies that protect our rights. Our mindset will play a key role in our successfully adapting to Industry 4.0.

Chapter 7
The Impact of the Industrial Revolution on Our Lives

Technology and the industrial revolutions have had a profound impact on humanity. The First Industrial Revolution deeply affected three different strata of society: the aristocracy, the clergy, and the peasantry.

Aristocrats were noblemen who inherited their titles and wealth from land ownership. They were exempt from paying taxes and were allowed to collect dues from the peasants who farmed their land.

The clergy were all the people who were ordained for religious duties, including priests, monks, and nuns. The clergy were divided into a hierarchy, with the Pope at the top and parish priests at the bottom, all of whom originally had to be members of the nobility.

The peasants, or "common people," constituted by far the largest percentage of the population.

This hierarchical structure gradually eroded over time. With the advent of technology, merchants began to earn great fortunes, which ultimately weakened the domination of the aristocracy. When the First Industrial Revolution took hold, entrepreneurs and industrialists caused a great shift in the balance of power. Initially, investors and industrialists alike came from noble families. Historically, they were the only ones who possessed enough capital to finance new inventions. Ironically, most aristocrats had little interest in industrial innovations. They viewed such investments mostly as a gamble (History, 2020).

In time, entrepreneurs without a noble background began to emerge. By the early nineteenth century, the wealth of successful industrialists far exceeded that of land-owning aristocrats. The once undisputed power held by noble landowners gave way to businessmen and moguls of industry. Uppermost in this new hierarchy were powerful industrialists who presided over global commercial empires.

The forces of growing industries and the changing forms of commerce transformed social conventions. For example, societal mores regarding the clergy changed significantly after the First Industrial Revolution. For centuries, the relationship between science and religion had been problematic. The clergy had long held a unique position in society as the arbiters of truth. For this reason, the clergy were regarded as only one step below the aristocracy. Over time, however, their prestige as an essential part of society devolved.

Priests were once among the most highly educated members of society, and for centuries enjoyed an almost undisputed advantage among their parishioners. However, as science progressively unraveled the mysteries of the universe, the clergy's role in society gradually lost its importance, and the authority of the Church weakened.

In the eighteenth and nineteenth centuries, maintaining

factories and operating machinery put into motion an unprecedented expansion of public education. As a result, a new educated demographic emerged, which included managers, clerks, teachers, engineers, lawyers, and doctors, who became known as the "middle class."

The peasants were greatly impacted by the First Industrial Revolution. For generations, farmers had been at the lowest rung of society. Their livelihoods were largely determined by seasons, weather, and the dictates of the landowners whose property they lived on. Landowners not only had the right to decree how many laborers they needed, but they could also impose taxes, which theoretically were applied to the costs associated with the construction and repair of roads, dams, windmills, canals, and so on.

The First Industrial Revolution not only ushered in innovative methods of production, but it virtually altered all of agriculture. Farming no longer had a low status within the economy, and, in any case, with an alternative means of earning a living, many peasants began to migrate to cities in search of a better life. The displacement of rural populations in favor of cities accelerated urban growth.

A New Era of Social Disparity

There is no doubt that much of the success of the Industrial Revolutions was made possible by the drudgery of laborers in mines and factories across the globe. Miners and assembly line workers alike, not to mention child laborers, worked in wretched conditions. Yet, although they were invaluable to businesses, they were often degraded and looked down upon by their employers. Compared to workers in the mid-eighteenth century, workers in the late nineteenth and early twentieth centuries made better incomes. However, their benefits were marginal compared to the almost obscene profits enjoyed by the tycoons running those businesses.

By the turn of the twentieth century, most workers lived in appalling slums or tiny houses, worked 12-hour to 16-hour shifts, and subsisted on poor diets. By contrast, the wealthy industrialists who employed them often lived in opulent luxury, with many of them owning fabulous mansions, private railway cars, and luxurious automobiles. Despite the breathtaking economic growth of nations, the gap between the rich and the poor who worked for them grew ever wider. As the disparity worsened, workers began to stage strikes, but their protests were often put down violently and sometimes with lethal force.

New social theories addressing the class struggle began to spring up everywhere. For example, in 1840, the German economic philosopher Karl Marx began to expound his radical philosophical proposals to allow workers to own the means of production to form equal and just societies. More political theories like his followed, spawning the concepts of communism and socialism. All of those theories were based on the idea of allowing societies to take control of production in both industry and agriculture in order to spread wealth evenly.

Workers Started to Gain Benefits, But—

The worsening social disparities gave rise to political and social movements. Protests led to clashes between workers and business proprietors. The disruptions and social unrest led to the formation of "unions," which made it increasingly difficult for businesses to exploit their workers. The unions also made significant inroads for workers by shortening their workweeks to five days and their workdays to eight hours, ensuring safer working conditions, and providing benefits like health care, social security, and unemployment insurance.

Unfortunately, due to a lack of organization and governmental support, unions over time have declined in membership and have lost their initial power. That has left workers more or less "back on their own," particularly in the private sector. The figure below shows the decline in union membership in the private sector within the United States.

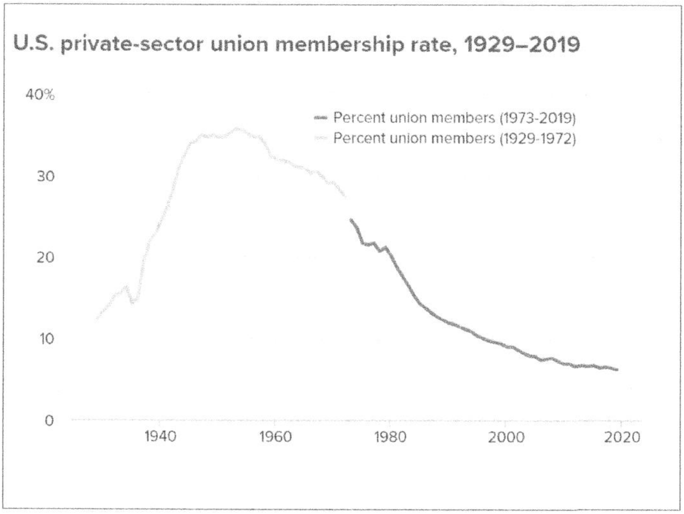

Source: Hirsch, 2008

The decline of unions directly impacted the middle class, from which a new economic class emerged, labeled "low-wage workers," who have no basic benefits such as pensions and health care, which had previously been common. The American journalist Harold Myerson warned that with the absence of powerful unions, the American middle class would shrink, and the concentration of wealth would increase. Under such conditions, corporate domination of the government inevitably grows.

The Service Sector in the Fourth Industrial Revolution

The arrival of the industrial age brought another important change in our modern way of life. The increase of manufacturing output, the growth of urbanization, the relentless rise of the world's population, and the expanding economies around the planet have brought about the "service sector." This segment of the workforce is not involved in production, but rather in providing services for those able to afford them. Although some of these jobs have existed for millennia—for example, innkeepers, doctors, and even lawyers—the arrival of the industrial age helped to grow and nurture this new sector.

For instance, the growth of nightlife has led to restaurants being open for longer hours, and the ease of travel has increased the demand for hotels. Such businesses require clerks, secretaries, managers, and so on. These same positions are also needed by governments. The new demand for all these jobs created a new job market, and entirely new professions began to spring up, such as flight attendants and computer programmers. These types of jobs have become the new backbone of the middle class.

It is interesting to note that the same industrial revolutions that helped to create so many new jobs are now taking them back. Based on fifteen years of data from the U.S. Bureau of Labor Statistics, a recent study concluded that the present-day industrial revolution has upended the service sector in much the same way the previous revolutions changed manufacturing in the nineteenth and twentieth centuries (Nath et al., 2020).

Initially, people who delivered services needed to be close to their customers. However, with the rise of the Internet, mobile devices, and ubiquitous access to data, that is no longer the case. Many jobs created by the industrial revolutions of the past have since been automated. Self-service, combined with automation and online access, has taken over many basic transactions and, in many cases, totally eliminated the need for human intervention.

In 2019, while visiting a shopping center in Las Vegas, I saw a crowd of amazed people standing in front of a machine that was busily animating its arms. When I got closer, I saw that it was a robotic bartender, mixing a drink based on any order placed on a touch screen. The mechanized arms mixed gin and vermouth and then added an olive as garnish. It was the perfect martini and hygienically void of any fingerprints—perhaps the cleanest drink one might ever get. Incidentally, there was no need to tip the robot, but it did express appreciation if you did. It was both polite and obedient and did not require time off, a monthly salary, health benefits, or a pension. If you were a bar owner, would you hire robots or human beings?

Source: https://www.dezeen.com

The Swing Riots and the Lessons Learned

The swing riots were a series of widespread uprisings that started in 1830. During the riots, protesters destroyed agricultural threshing machines, which were fast displacing human workers at the time. They also committed arson and targeted workhouses and silos (Norman, n.d.). The situation worsened when a group of skilled textile workers formed a secret organization, calling themselves "Luddites." The clandestine group continuously protested the labor-economizing technology because it posed a threat to their livelihood. The costly destruction of equipment eventually caused police to be stationed around the machines to guard them. It didn't stop there. On several occasions, the British army clashed with bands of marauding Luddites. At one point, more British soldiers were fighting the Luddites than soldiers fighting Napoleon on the Iberian Peninsula.

Source: https://commons.wikimedia.org

The riots were not necessarily motivated by fear of technology and automation. The rage boiled up from the massive loss of jobs caused by those advances and the resulting poverty that followed.

Innovation has always been essential to human survival. However, in many cases, people have only considered innovation from a technological perspective, but innovation is not just technical. It includes the revision of long-standing policies, management techniques, management styles, and organizational structures. The failure to adequately prepare for new technological advances will not only create conflict between managers and workers, but will negatively impact the whole society.

In the 1960s, one of the most popular series on American television was *The Twilight Zone*, which delved in fantasy, science fiction, and dystopian fiction. One episode dealt with a character named Mr. Whipple and his obsession with replacing his company's human staff with robots. In one scene, Whipple triumphantly introduces a timesaving machine, declaring, "This is the X-109B14, which eliminates sixty-one thousand jobs, eighty-one thousand needless man-hours, per eleven working days, and four million dollars each year in expenditures for employee hospitalization, employee insurance, employee welfare, and employee profit participation" (Road, 1964). The year this episode aired, the probability that an entire enterprise could be run with zero employees still seemed far-fetched. But today's technology has far outpaced that represented in 1964.

Today, if a business process is transmuted into a function performed by a robot or an algorithm, managers immediately let go of the employees who used to do the job, in order to save costs and increase profits. However, we have a huge opportunity to leverage the collective intelligence between human minds and the artificial intelligence of robots and algorithms. Computers can deal with highly complex data, produce unerring results, and function in harsh environments far better and faster than humans can. The human brain, on the other hand, is a source of limitless creativity, capable of conceiving and visualizing an infinite variety of new concepts. In terms of medicine and health care alone, the human voice and human touch have healing powers in themselves, and only humans can respond appropriately to human emotions. If artificial intelligence and human beings collaborate with each other more compatibly than they do now, the union of the two will surpass the capabilities of any number of human beings or computers alone.

Let's consider the following scenario. A radiologist specializes in reading X-rays, CT scans, MRIs, mammograms, and other electronic images, and earns an average annual salary of $400,000. A radiologist's core function is to analyze these images and then send the results to the physician who requested the scans and now can discuss them with the patient. According to Rosetta (2021), the average waiting time for requesting physicians to receive non-urgent radiological reports is:

Two or more weeks for MRIs
Two weeks for mammograms
One to two days for X-rays
One day for CT Scans
No wait for ultrasound images

The waiting time is divided into three stages: (a) the time for the radiologist to get the images from the machines; (b) the time it takes the radiologist to analyze the imagery and write a report about it; and (c) the time that it takes the radiologist to send the report back to the requesting physician. The first and last of these stages are virtually instantaneous. Most of the time is currently consumed in the analysis of the imagery and the writing of the report. That middle stage can vary in length, based on the number of patients whom the radiologist sees every day, and the complexity of the patients' conditions.

Given advances in artificial intelligence, image recognition, and machine learning algorithms, a radiologist's diagnosis will be automated in the future by the machines themselves, which will have the ability to read the images, generate diagnostic reports, and send the results directly to the requesting physician in real time without any human intervention. This process will lead to cheaper, faster, and more accurate diagnoses. The reduction in expenses will come mostly from not having to pay the radiologist's salary.

Dr. Robert Schier, a neuroradiologist with RadNet Northern California, stated, "I guess that in ten to twenty years, most imaging studies will be read only by machines. The results will be transmitted directly to the referring physician without input from a human radiologist" (Guilford-Blake, 2020). Although Dr. Schier has projected ten to twenty years, I believe it will be far less than that. For one thing, the pandemic has given many hospitals no choice other than to automate many of their processes just to keep up with the massive number of patients affected by the Covid crisis. I now often have video appointments with my doctor, which I did not have before the pandemic.

The digital transformation of the radiologist's job does not mean that there will be no need for radiologists in the future. Radiologists will be trained to enhance the performance of algorithms, to improve robotics to produce more accurate results, and to monitor radiological equipment to ensure that it is working properly. Software engineers cannot do that job because they do not have the know-how to decipher the images, much less to enhance the analysis of the images. For these reasons, medical schools need to alter their curricula to incorporate software engineering as part of their students' training.

Collective intelligence between human beings and

computers will always achieve more than the singular efforts by each one of them. The "wisdom of the crowd" has been acknowledged as far back as ancient Greece. For example, Aristotle noted in *CATEGORIES* (Aristotle 2016) that a gathering of many unremarkable men can often come to a better collective judgment than the conclusions of one great individual. That still applies today, except that artificial intelligence has been added to the equation.

Reconsidering the scenario of robotic bartenders, that is not going to wait for ten or twenty years. It's already here. In bartending school, students learn about bar equipment, bar setups, liquor types, beer keg sizes, sobriety, and other matters. All of that is not going to be replaced by robots or software engineers. After robots are up and running, they are always limited by their programming, and there will always be situations when they need to be reprogrammed or updated for new drink recipes. Bartenders will always be needed to supervise, train, and manage any robotic programming. Therefore, bartenders will not lose their jobs because of digitalization. Instead, their job descriptions will be defined in a more sophisticated way.

Unfortunately, no one has trained bartenders or anyone else in similar situations about the new changes in their careers. Such training is not available because the private sector does not feel obliged to train workers to catch up with new technology. After all, it is easier and cheaper to simply lay people off and hire others who already have the needed skills and knowledge. Also, in most cases, workers today do not have unions to fight for their rights. Furthermore, government policymakers are not technologically mature enough to see how Industry 4.0 is reshaping social behavior and income distribution. Did we learn anything from the swing riots?

In my observation, both citizens and communities,

especially those in the service sector, have been left in the dark about how all-encompassing the Fourth Industrial Revolution is. Today, people have no option other than to take responsibility for their own career development and to emotionally and financially cope with this overwhelming and baffling transition we are all a part of.

Chapter 8
It's All About Our Mindset

Human Intelligence

Human intelligence enables us to learn from experience, to adapt to new situations, to understand abstract concepts, and to use this knowledge for our survival. Colom et al. (2010), who are scientists at the National Institutes of Health, defined intelligence as "a general mental ability for reasoning, problem-solving, and learning."

Twelve thousand years ago, when the Agricultural Revolution began, it became possible to feed thousands of people—not only the inhabitants of crowded cities, but also legions of soldiers deployed in the field. The scale of such agricultural yields was brought about by technological innovations. But the improved husbandry brought new challenges with it. As food production grew, the population expanded, which then increased the demand for food. The early farmers relied solely on their brainpower to process their production data.

Our ancestors believed that they had to build temples, make festivals, offer sacrifices, and allocate portions of their land to God. For instance, 6,000 years ago, in the cities of ancient Sumer, which is now the southern part of Iraq, temples were not only used for religious worship, but also served as political and economic centers. In Sumerian life, God owned land, paid salaries, and built dams and canals (Harari, 2017).

God did not die or have children to inherit divine wealth. As economies grew and wealth accumulated, the distribution of that wealth to pay salaries and to extend and collect loans became more than the priests could cope with, because it was too difficult to remember which estates owed what, or which employees had received their salaries, or which ones had failed to pay God's tax, and so on. There needed to be a way to create order out of all this chaos.

Historians credit the Sumerians with inventing the first writing system and the first mint coinage, some 3,500 years ago. Our entire history as a species would be unimaginable without either a written form of communication or a monetary exchange system.

The ancient Sumerians had to constantly innovate to keep up with the changing times. They invented and also improved a wide range of technology, including the wheel, arithmetic, geometry, irrigation, and various tools.

If we apply the definition of intelligence here, the Sumerians used their mental ability for reasoning and solving the inherent problems brought on by great wealth. By inventing coinage and a writing system, they harnessed their expanding wealth while adapting to change.

Sumerian priests had to learn, adapt, and change their mindset to embrace reading and writing. Otherwise, others could and would have replaced them. We are living in an age today that requires us to use the gift of our natural intelligence to learn, change, adapt, and grow for much the same reason: to keep pace with all of the technological advances upon us. What was technologically advanced yesterday will not be technologically advanced tomorrow.

I know that most of the technical information in this book will be outdated within the next five years, and that is part of modern life. Change itself will inspire me to write yet another book, or at least a new edition of this book.

In *The Technology Fallacy*, the authors describe several surveys they conducted to measure the ongoing digital disruptions within organizations (Kane et al., 2019). One of the surveys asked 3,300 respondents what they considered to be the most important skills that employees should have today to succeed in a digital workplace. Forty percent of the respondents reported that being change-oriented was the most important skill for employees to possess. Other associated characteristics that they cited were: adaptability, flexibility, agility, and innovation, none of which are possible without changing our whole mindset.

Fixed Versus Growth Mindsets

"He's naturally intelligent!"

"She's brilliant!"

"He's very smart!"

These are things we might say to describe people based on their intelligence. Exactly what makes some of us more intelligent than others? Many philosophers and psychologists have been fascinated by the concept of intelligence and its effect on our ability to learn, to solve problems, and to succeed in life.

Many people correlate intelligence with the level of one's score on an IQ test. That score has also been used to predict academic and professional success. In recent years, however, traditional thinking about intelligence has been challenged by new and far-reaching concepts.

Dr. Carol S. Dweck (2007), a psychologist and professor of psychology at Stanford University, has described two different types of mindsets, which she calls "fixed" and "growth." Those mindsets, she contends, play a much bigger role in our success than our inborn intelligence.

Individuals with a fixed mindset believe that intelligence, talent, personality, and other traits are static. In other words, either you have it, or you don't. We tend to hear this when people say:

> I've always been good at math.
> I already know everything I need to know.
> I'm too old to learn how to do that.
> I've always struggled with technology.

High achievers with this type of mindset have a recurring urge to prove themselves over and over again. They believe that if they have a finite amount of intelligence, moral character, talent, or whatever, they need to demonstrate that in every situation. By this ethos, individuals strive for success in their given field at any cost. Falling short of that undermines their identity as smart, skilled, intelligent, or suchlike.

This is a trap that all of us can get into consciously or unconsciously. For instance, in one of my prior jobs, there was a mission-critical system that had been designed some twenty years before I got there. The team that put the system together had broken new ground by connecting 500 different computer servers. Prior to that, it was unprecedented that thousands of pages of bulletins could be processed in a short amount of time. At the time, the innovation was a brilliant coup because it exponentially increased the processing power of the system's output. The team members received accolades for their work and continued augmenting the system with more features for almost a quarter of a century.

In the past fifteen years, however, the concept of "cloud computing" has become popular. The "cloud" replaced the need for dedicated servers by simply providing the needed storage and processing power as a software service. Users are now able to avail themselves of this convenience whenever and wherever they require it. These newer and more economical technologies have rendered the old systems obsolete and too costly to maintain. Yet, the very team that once came up with groundbreaking solutions was slow to accept those facts. The team's denial represented a collective fixed mindset. Hearing and seeing reality through such a filter prevented them from accepting the obvious.

I remember hearing members of the team say: "No other system can do the same job!" or "Our technology is so advanced that our competitors can never build anything like it!"

It was obvious to me that I was listening to the musings of a "fixed mindset," and not ideas based on reality. With such thinking, arguments are advanced that continuously rationalize what cannot be rationalized. Futile attempts at succeeding one more time are repeated over and again. Historically, this frame of mind has kept entire peoples blinded to the truth. The situation is not unlike that of the cave dwellers in Plato's "Allegory of the Cave." Those cave dwellers were a tribe who grew to believe only in the shadows they saw in their cave. When one errant member discovered the outside world and returned to inform them of it, they threatened his life, believing that his "truth" was a form of heresy, abhorrent to their way of thinking.

In my experience, I sensed that what was really behind the team members' resistance was their dread of having to undergo yet another learning process.

Fixed mindsets can be observed on different levels. They can be found in individuals, teams, organizations, companies, and even entire nations.

Alternatively, a growth mindset maintains that intelligence, talent, personality, and other traits and capabilities can be developed and increased. They are not static or predetermined, but focus on processes and outcomes. People with this mindset believe that their basic abilities can be improved. They readily adapt to the "how" and "why" of any changes that need to be learned, negotiating each new stepping-stone toward further development and improvement. In the end, their basic qualities are cultivatable and flexible. These virtues, when coupled with time, dedication, and commitment, not only expand the intelligence of such persons, but also improve their ability to continuously adapt to new ideas.

Clear examples of fixed and growth mindsets on a corporate level would be Walmart and Amazon, respectively. Walmart has been in the retail business since 1964. Amazon came onto the scene in 1994 as an online bookseller. From its inception, Amazon has taken on new and innovative approaches that have continuously placed it ahead of its competitors. The expansion of its online business has included a vast array of goods and services that has prompted its move to warehouses and distribution centers, or "brick and mortar" spaces.

Some of Amazon's most impressive strides forward have included developing Amazon Web Services as well as the entire concept of Cloud computing. Amazon Web Services alone have become one of the primary components of the e-commerce giant's business portfolio, generating more than $12 billion in annual profits.

By contrast, Walmart did not change its classical business model until recently. Throughout the years, it has stayed focused on delivering retail services to its customers within physical locations. When the two companies are placed side by side, it is clear that Walmart operates from a fixed mindset, whereas Amazon operates from a growth mindset. The chart below represents the variables of each company's stock value from 2016 to 2020. Amazon's meteoric rise contrasts sharply with Walmart's flat line.

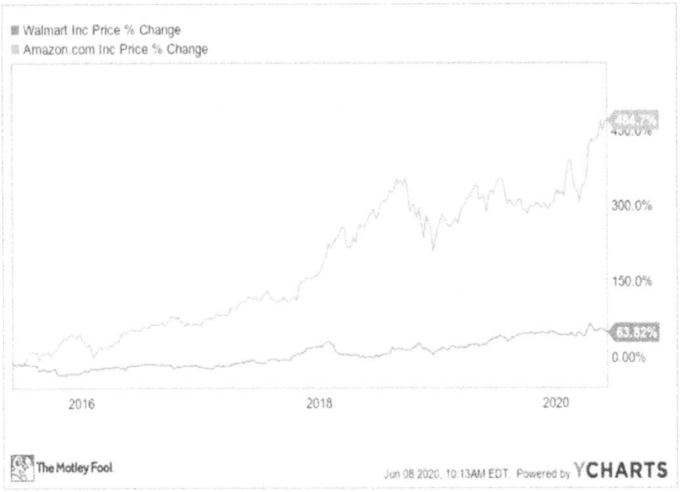

Source: https://www.nasdaq.com

All of us have a mixture of fixed and growth mindsets, depending on the subject, and we can always change our mindsets when we become aware of them. Carol Dweck (2007) has stated:

> Everyone is a mixture of fixed and growth mindsets. You could have a predominant growth mindset in one area, but there can still be things that trigger you into a fixed mindset trait.... So I think we all, students and adults, have to look for our fixed-mindset triggers and understand when we are falling into that mindset.

Fixed mindsets exist in all of us to varying degrees, so it is important that we become aware of them. If we recognize them within ourselves, we can work on changing them. Our inherent mindsets make up our convictions. Staying flexible and not allowing any set of fixed ideas to calcify can be challenging at any age, but particularly as we get older.

"You can't teach an old dog new tricks" is a well-known truism about human aging. The older we get, the more we tend to get set in our ways. To see that, one need only compare an older person's mental agility with a child's. In our youth, our minds are at their nimblest, which is why, for example, children can learn languages far faster than adults can. However, our growth mindsets enable us to realize that learning itself is an indispensable tool for energizing our minds. Learning improves our understanding and helps us to better cope with the constant changes that life brings on. Like water flowing over rocks, continuously exposing ourselves to new ideas can rewire our thinking and wear down even the most rigid of our beliefs.

A Digital Mindset Is a Growth Mindset

The differences between the two kinds of mindset are essential to success in Industrial Revolution 4.0. Cultivating a growth mindset is critical for developing the skills and knowledge necessary to work effectively and efficiently in this rapidly changing world. Individuals who accept and adjust to change, who learn from their mistakes, and who can accept criticism will continue to improve. That said, it is not an easy journey for most of us, because adapting to new technological advances is a struggle.

Hidayatno et al. (2019), researchers at the Department of Industrial Engineering at the University of Indonesia, measured and assessed employees' use and acceptance of technology-based products and services in an automotive company that was undergoing a digital transformation as it converted its factory into a "smart factory." To meet its production needs, the factory had been operating on a 24-hour basis. Each time one of its "old-school" machines broke down led to significant financial losses for the company.

To solve this problem, the researchers built a monitoring system that transferred data from the factory floor to the managerial office in real time. The goal was to streamline as much data about the operating machines as possible through a network of Internet of Things sensors in order to get a better read on the status of the machines. The new Internet of Things sensors were able to predict potential failures before they happened, which eliminated the need for workers to constantly check on them. That, of course, significantly cut back on man-hours.

The researchers concluded that, from a technological standpoint, the new system worked well. However, the research revealed a downside, which was that the workers did not totally accept the system. To overcome that resistance, the researchers suggested reconstructing the organizational hierarchy of the factory, since some new roles and responsibilities had not existed prior to the digital transformation.

§The researchers also recommended helping the workers to use and adapt to the new technologies. That is not surprising, since some of the workers had been at their jobs for years and were set in their ways. In other words, their mindset had become permanently fixed, causing them to resist everything that was new.

This is only one of many examples of how staying firmly rooted in our habits does not serve us. Incorporating new technology into the stream of life has always required mental and emotional flexibility, because change is a double-edged sword. Along with solving problems, it also stirs up uncomfortable feelings and insecurities.

We all have a part of us that resists change. That is part of the human condition. Unfortunately, that resistance does not serve us well in the changing times we live in today. Keeping one foot in the past only weakens workers' chances of future employment, because when their employer moves forward with a digital transformation, those who are unable to "keep up" will be cut.

Capitalism Will Not Save My Job

In capitalism, private enterprise controls every facet of production, which includes land, labor, and capital. Private companies control the deployment of a mix of these factors to increase their profits and their efficiency. They use their profits to earn additional profits. Based on this fact, companies are always going to use the most efficient means and technologies to increase their profits, regardless of the impacts on their workers. That is sad, but a reality.

Increases in profits have often been associated with losses of jobs. Economists have often accurately forecast future job losses by observing the economic conditions of their time. For instance, in the 1930s, when John Maynard Keynes warned of "a new disease..., namely technological unemployment," many economists, government officials, legislators, and others proposed initiatives to address it.

In 1928, the Republican Party declared in its national convention platform, "We realize that there are certain industries which cannot now successfully compete with foreign producers because of lower foreign wages and a lower cost of living abroad" (Peters & Woolley, 1928). The threat of the Second Industrial Revolution was the cheap labor available offshore.

In 1961, President Kennedy sent a message to the U.S. Congress, in which he said, "Large-scale unemployment during a recession is bad enough, but large-scale unemployment during a period of prosperity would be intolerable" (Kremen, 1974). He then created the Office of Automation and Manpower within the Department of Labor, identifying "the major domestic challenge of the '60s: to maintain full employment at a time when automation, of course, is replacing men." The threat of the Third Industrial Revolution was automating physical labor.

In 2011, during an interview on NBC-TV, President Obama suggested that technology and automation were in part responsible for the U.S. economy's sluggish job growth.

"There are," he said, "some structural issues with our economy where a lot of businesses have learned to become much more efficient with a lot fewer workers" (Curry, 2021). That is precisely what fuels the engines of capitalism: using any possible means to increase profits. The threat of the Fourth Industrial Revolution is automating intellectual labor, which includes integrating and automating as many business processes as possible.

The warning that Keynes issued in the 1930s pointed to technology as the cause of unemployment. But the technology of his time was not nearly as advanced as what we have today. Nowadays, Industry 4.0 has made it possible for machines to run on their own, using artificial intelligence to make decisions without any human intervention. In many cases, there will not even be a need to search for cheaper labor overseas because there will be no need for labor at all in the first place. Take, for example, JD.com, a Chinese e-commerce company that can fulfill 200,000 orders a day for products and services, but employs only four people. Most of the old types of jobs have been replaced by computers or robots (Rajamanickam, 2019). The new digital industries generate far more revenue than most of the traditional industries did, and with far less manpower.

As of today, there is no clear government strategy that addresses today's technological advances and their impacts on jobs, simply because technology is moving faster than policies and legislation can be made into law. Not surprisingly, as I mentioned in Chapter 6, the chair of the Technology and Innovation Commission for the city of Long Beach stated that she had no background in technology! How can someone who does not know the latest trends in technology put forth policies that affect the impact of technological advances on people?

The problem is so serious that it compelled Robert Reich (2015), the former U.S. Secretary of Labor and currently a professor of public policy at the University of California at Berkeley, to pose the question, "Why is the economic system that made America strong suddenly failing us, and how can it be fixed?" That problem is bigger than I can solve in this chapter. It needs to have a complete book to cover it. But one thing we know for sure: technology is not the problem. Rather, technology is a tool that has given us the ability to improve the quality of our lives.

Meanwhile, the question that most needs to be answered is: How can an average citizen cope with and survive the digital transformation that has come with Industry 4.0 until the economic system and public policies are fixed?

A Growth Mindset and Strategic Thinking Will Save Us

In October 2012, I accepted a job offer from The Boeing Company to work in its office in Maryland, so I had to relocate to the east coast from Seattle. However, the job came with a relocation package, which made the move a bit easier.

Two weeks later, I landed in Virginia for a few days of training. Then, the following week, I drove to Maryland to join my new team. After living in a hotel for two weeks,

I found an affordable apartment near my workplace. For the most part, things went well for the next five months. But then, without warning, all the members of my team, including me, received a two months' notice, informing us that we were being laid off because our statement of work had been outsourced to the Corporation Service Company.

Needless to say, I was shocked! I had just moved from one coast to the other, and five months later I was out of a job! When I asked my manager to explain the logic behind hiring me, relocating me, and then laying me off in such a short time, she said that she had been laid off herself! I felt that the whole situation was entirely unethical.

Now, nine years later, when I think about it, I see the situation from an entirely different perspective. If I had been aware of the company's strategy back then, I would have realized that my job was not aligned with Boeing's strategic direction. For instance, the company had a strategy of outsourcing any possible statement of work that could reduce costs, so it could better compete in the marketplace. Given what I know now, it makes perfect sense that Boeing outsourced my job. I was part of a 24/7 network operation center, which worked three shifts a day, monitoring problematic network incidents. However, the tools and technologies we were using were old, and many of the monitoring tasks could have been automated, which would have greatly reduced overhead.

In a perfect world, all workers should think strategically, so they can accurately gauge the lifetime of their jobs, given the rapid-fire technological advances that are occurring all the time. That would make the relevancy of jobs within the next five to ten years much clearer. Rather than relying on one's employer or the government for job protection, we must be informed about and prepared for the radical shifts of the current industrial revolution.

In my neighborhood, the local grocery store used to employ eight cashiers. There were only four self-checkout stations, with one employee assisting all the customers who used them. Essentially, that employee was training the customers in how to use the self-checkout machines. After enough customers were trained, the store decreased the number of human cashiers. Within months, the store added two more self-checkout stations and closed two of the human-operated ones.

When I walked in one day, I saw long lines of people at all of the human-operated stations. Not wanting to wait fifteen or twenty minutes to check out, I opted for one of the self-checkout stations, and was back in my car five minutes later. Was that upsetting for the employees who lost their jobs? Obviously, the managers of the store were not upset. After all, they and many other retailers were now competing with Amazon, which provides cheaper and faster products for its customers and does not have to provide or maintain self-checkout machines, pay many salaries, purchase much insurance, and so on.

If I worked as a cashier, I would take the hint and go back to school to get some training to prepare myself for some other kind of work. That would give me the advantage of being one step ahead of my employer. Leaving one job as a cashier for another job as a cashier somewhere else would not make sense, because I could not guarantee that the next store would *not* automate my job out of existence. I am not claiming that the current situation is either good or bad. I am just acknowledging reality. That is how these systems work. All of us must use our growth mindset by practicing the kind of strategic thinking needed to adapt to the technological tsunamis currently overtaking our planet.

Charles Darwin (University of Cambridge, 2016) observed that "it is not the strongest of the species that survives, nor the most intelligent that survives. It is the one that is most adaptable to change."

Chapter 9
Privacy in the Digital Information Age

According to the United Nations Population Fund more than half of the world's population currently inhabits urban areas, and that figure will likely increase to 68 percent by 2050 (Elmjid, 2017).

Centuries ago, the majority of the world's population lived in rural districts, but the tendency of humans to live near each other has steadily gathered momentum. Today, several cities exceed population figures of more than 20,000,000 inhabitants (worldpopulationreview, 2021). With the population numbers having reached a critical mass, there is now a voracious demand for services that are innovative and fast.

In Industrial Revolution 4.0, all of our technological infrastructures have been entirely remapped and reconfigured, mushrooming so quickly that an individual anywhere on the planet now has access to information of every kind. With a myriad of services available and unprecedented access to history's greatest knowledge base, one would think we would be living in a kind of Utopia.

In Chapter 3, I touched on the issue of how Industry 4.0 has increased agriculture's capacity to meet the demands of the swelling urban populations. In Chapters 4, 5, and 6, I underscored the need for smart city initiatives to resolve the daily occurrence of problems related to parking, traffic, and street size.

Numerous and wondrous as these technological advances have been, they have also put into motion a rush of intellectual confusion, ethical quandaries, and emotional challenges that had not existed before. In my opinion, one of the principal issues in this technological maelstrom is that of data security and privacy. In terms of importance, I would single out these issues as second only to global warming.

Security and Privacy Were Not Part of Designing the Internet

In 1958, President Eisenhower authorized the creation of the Advanced Research Projects Agency, or ARPA, whose purpose was to catch up with Sputnik, the first satellite sent into space. Created by the Soviet Union, Sputnik bore the distinction of being the first manmade craft to complete a low-Earth orbit (Kremen, 1974).

U.S. military forces recognized that the historic Soviet launch had serious implications for the national security of the United States, because the Soviets had not only placed a satellite in space, but they had also developed a potential military advantage. The United States responded by investing heavily in ARPA, which ultimately became the Defense Advanced Research Projects Agency, or DARPA. Scientists from various universities were teamed up to work on the initiative, one of whose most crucial issues was: What would happen if the Soviets attacked the nation's TV and telephone network lines with a missile? In the event of such a disaster, how would government officials communicate with each other?

In 1962, Joseph Licklider, a scientist from the Massachusetts Institute of Technology, proposed the first version of a solution to the problem, which he called the Intergalactic Network, which consisted of a link of two computers that could talk to each other. This remarkable new alternative means of communication was the breakthrough that the United States had been searching for (Robnett, 2020).

That innovation remained in use until 1969, when DARPA delivered its first messages between two remote computers, one operated at UCLA in Southern California, and the other operated 300 miles away at Stanford University in Northern California. Each of the mainframe computers was the size of a small house!

In 1971, ALOHA Net at the University of Hawaii joined that network. Two years later, London's University College and the Royal Radar Establishment in Norway also joined in. The interconnected computers made it possible to share massive amounts of data instantly, not only across the United States, but also between the United States and its allies abroad. That international network marked the birth of what we now call the Internet.

Designed as a means of communication for government, military, and research purposes, the Internet was not originally intended for public use. Hence, many architectural considerations were missed or overlooked at the time of its conception. For example, for any computer to connect today to the Internet or to any sort of network, it needs to have an Internet Protocol (IP) address, which is a unique code that identifies individual computers, phones, and other devices. The original maximum number of such addresses was less than 4.3

billion. At the time, that number was considered to be more than enough for both current and future uses. However, in the late 1980s, the Internet experienced a dramatic growth, which soon led to the depletion of all the unallocated IP addresses. Several solutions to that problem had to be devised to overcome the shortage, but the original architecture of the Internet remained unchanged.

One of the glaring flaws of today's Internet is its significant lack of security and privacy. Again, the main focus of the Internet in the 1970s was to serve as a means of "connection." The idea of having two computers talking to each other followed by a network of computers in remote locations was a dream come true. Security at that time was thought of mostly in physical terms, such as preventing someone with programming knowledge from accessing those computers. No one back then saw the Internet as being an essential part of everyday life.

As of today, according to zone-h.com (2021), there is at least one Internet attack on computers somewhere in the world every single minute. The tracking done by zone-h.com does not, of course, include attacks that remain "untraceable" or that are lying dormant.

Even at this late date, many computer scientists have urged that the Internet be redesigned to overcome these security and privacy issues. But such a proposition is no longer feasible, since almost the entire global population, all commercial businesses, and all government agencies can no longer function without the Internet. The long and the short of the situation is that we have to live with the Internet as it is, becoming ever vigilant against cyberattacks and the violation of our privacy.

Data Privacy and Data Security

Data security refers to the protection of digital information from any unauthorized access, corruption, or theft throughout its entire lifecycle. Data security encompasses all aspects of hardware and software security.

Data privacy, however, is the right of individuals to have control over how their personal data is collected and used. The challenge today is that many people have lost control over their data and don't even know where or how it is being used. The truth is that information about each of us is being collected every single moment of our lives. For instance, most of us use social media and enjoy the free services that platforms like Facebook, Twitter, and Instagram offer us.

Clario.co (May, 2021), which surveyed various apps to determine the amount of data that those apps were collecting from their users, discovered that the collected personal data included the users' names, ages, genders, sexual orientations, races, religions, home addresses, live locations, phone numbers, and so on.

At one time, only governmental agencies held such data on citizens. Nowadays, private companies like Facebook, Twitter, and TikTok have far more data about individuals than any governmental agencies do.

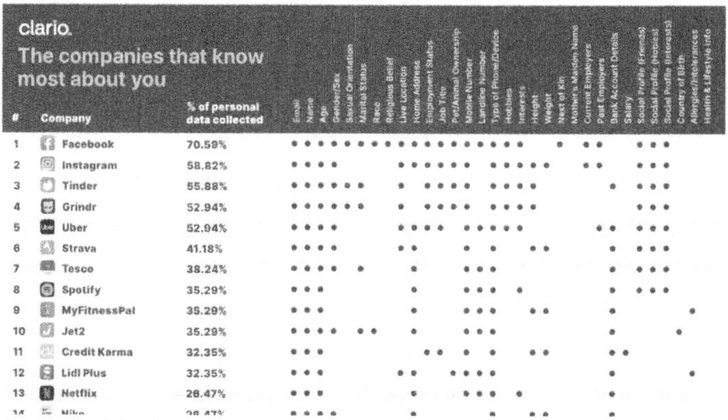

Source: slynchuk, 2021

Within the realm of Industry 4.0, data is the new gold that everyone is clamoring for. Many companies are doing everything they can to get users' data so that they can make sales based on it. Collecting and selling that data is a multi-billion-dollar industry, filled with data brokers, advertisers, databanks, and so on. Companies are making deals and partnerships based on this data, while users' privacy gets breached along the way.

Breaches of data privacy can also have political consequences. For example, the Facebook–Cambridge Analytica data scandal came to the public's attention in 2018. That infamous case involved a British consulting firm, Cambridge Analytica, which had been able to harvest the personal data of millions of Facebook users. The company had created a personality app test that consisted of a series of questions that could build psychological profiles on Facebook users to determine who and what they liked and didn't like, as well as other information about them. Those questions were derived from a list of questions created by a psychological researcher at Cambridge University's Psychometrics Center.

Because 300,000 Facebook users took the test, Cambridge Analytica was able to access not only the users' data, but also data about their friends and the friends of their friends. All told, the numbers of users whose privacy was invaded exceeded 87 million. Cambridge Analytica used all the latest artificial intelligence techniques to create digital profiles for all the users to understand their behavior. For example, by gauging a user's algorithm, the app could determine if the user was a registered Republican or Democrat. It could even ascertain if users were still on the fence, based on their geographical location and digital profile.

Of primary interest to Cambridge Analytica were those users who were still undecided about which candidate to vote for in the 2016 presidential election. Once the programmers of Cambridge Analytica amassed that target group, they began to direct all sorts of political ads at them to attempt to sway them to vote for a particular candidate.

Because there were no legal restrictions on the privacy of the data, Cambridge Analytica committed no security violations. To the alarm of many, however, the clandestine operation had taken place with the knowledge of Facebook's executives, but without the knowledge of its users.

Facebook CEO Mark Zuckerberg first apologized on CNN for the situation with Cambridge Analytica, referring to the whole matter as an "issue," a "mistake," and a "breach of trust" (Segall, 2018). In April 2018, the Judiciary Committee in the U.S. Senate decided to interview Mr. Zuckerberg to investigate different data privacy issues, such as the Cambridge Analytica scandal, the whole phenomenon of fake news, and Russian meddling in U.S. elections. At the outset of the hearing, Mr. Zuckerberg wore a grave and worried look on his face.

Yet, by the close of his first day before the committee, his demeanor bore an air of confidence. Considering the seriousness of the inquiry that had been put before him, he must have been greatly relieved. The line of questioning came from "tech-challenged" senators who were patently ignorant of the menacing capabilities of social media, as became embarrassingly obvious from some of the questions they asked:

> Is Twitter the same as what you do?
>
> How do you sustain a business model in which users don't pay for your service?
>
> If I'm emailing within WhatsApp..., does

> that inform your advertisers?
>
> When you..., when you say "pipes," you mean...?

The entire senatorial procedure was woefully inadequate. Anyone who follows social media mocked the senators, blaming their sophomoric questions on their age, since some were in their late seventies or even eighties. People began posting humorous tweets like:

> Mr. Zuckerberg..., hypothetically, if someone's video cassette recorder won't stop flashing twelve, how would you suggest they fix that?
>
> Facebook's stock is up 4.5 percent on news that America is run by people who still own VCRs.
>
> If there was ever a time for senators to have an earpiece, it was today.
>
> Senator: My aides have given me this complex multi-part question to read to you in a halting and uncertain voice.
> Zuckerberg: Oh, no. Don't worry about that. Our new motto is, "We fixed it."
> Senator: That sounds wrong, but I don't know what to ask.

I don't find any humor in these taunts. The senators'

inability to reprimand the social media's dastardly behavior was a complete debacle, and I reject the premise that it had anything to do with the senators' ages. The plain fact is that they were just not up to date on the technological advances of our time.

If we go back twenty years, there were no Smartphones and no social media. In fact, there was no pressing need for politicians to be technologically literate, because there was nothing like today's technology to impact people's lives back then.

Technology has advanced so rapidly that businesses are now manipulating and taking advantage of people without any legal consequences. That is so because policies necessary to safeguard the public against unethical practices cannot be enacted fast enough to keep up with the technological changes. Thus, politicians with no technological background cannot be expected to legislate issues that involve, for example, cryptocurrency, autonomous vehicles, data breaches, and so on.

Technologically illiterate senators and other politicians are in the same position as the city officials in San Diego and Long Beach whom I described in Chapters 4, 5, and 6, whose inactions have led to citizens' problems with finding parking spaces, avoiding traffic jams, breathing unpolluted air, and so on.

The U.S. Congress, like all legislative bodies, needs to be one step ahead of technology in order to appropriately enact laws and regulations to protect the security and privacy of information. In the terms addressed in this book, legislators must be made aware of the degree to which Industry 4.0 has already impacted and is still impacting our economy as a whole, and employment in particular.

Battle of the Giants

Facebook was originally created as a social network for students at Harvard to connect. Over the years, many would argue that it has morally and ethically devolved into a platform for hate speech and fake news. I am not denying that Facebook and other social network outlets have connected people. However, it would be wise to more keenly observe Facebook and any other social media platforms in the same manner as with car manufacturers when mandates for seat belts were put in place. A lack of political leadership and clearly defined policies will surrender all authority to the multinational corporations. Without stewardship, they will be left unguarded and free to define policies that only benefit their own interests.

In January 2021, in a quarterly earnings statement for Facebook, Mark Zuckerberg declared, "We increasingly see Apple as one of our biggest competitors." He went on to accuse Apple of using a "dominant platform position" to push its apps while interfering with Facebook. "Apple," he said, "may frame this as a privacy service to its customers, but it's only in Apple's own best anti-competitive interests" (Morrison, 2021).

In response, Apple's CEO, Tim Cook, made a speech at the Computers, Privacy, and Data Protection Conference, countering Zuckerberg with the following remarks:

> Technology does not need vast troves of personal data, stitched together across dozens of websites and apps, in order to succeed. Advertising existed and thrived for decades without it. And we're here today because the path of least resistance is rarely the path of wisdom. If a business is built on misleading users, on data exploitation, on choices that are

> no choices at all, then it does not deserve our praise. It deserves reform. (Rogers, 2021)

So, who is right and who is wrong? Are Apple's strategies for the benefit of the public or for its own agenda, and why are these giants fighting over our private data? Doesn't data privacy refer to the rights of individuals to have control over their own data? Who is protecting the citizens' data privacy?

Current policies are not adequately regulating the technological advances that we have today, which is why the response to and reprimand of Cambridge Analytica was appallingly inadequate. If our governmental policymakers had been more technologically savvy and proactive, such scenarios would never have occurred. This ignorance within our community leadership of what is occurring all around us is not only affecting people's privacy, but also adding to the public's distrust of governmental representatives and public figures at large.

For example, Bill Gates has become a principal target of various conspiracy theories that have spawned during the spread of Covid-19. There is even a rumor that he created the virus so that he could then implant microchips in our bodies with vaccines. Supposedly, the vaccines would then allow him to track citizens and conduct experiments on them remotely. Then he could raise people's blood pressure or any other measurement to enact social genocide on portions of the population. A poll of 1,640 people revealed that 28 percent of Americans actually believe this theory (Carmichael, 2020).

There is no basis for the rumor, and Bill Gates has vehemently denied it in numerous interviews. But such wild allegations should give us pause. We should all take a step back and think about why anyone would circulate such a rumor? Is it partially because we are now aware that giant tech companies are hijacking our privacy? Or is it because many people now believe that no one cares about their privacy? Or perhaps the public at large has just lost trust in the system altogether?

Are You Finding a Parking Spot Or Giving Up Your Privacy?

We all enjoy the online services provided by such technology companies as Google, Facebook, and Apple. However, what are we sacrificing for these "free" services?

In Chapter 5, I explained how severely the nationwide parking problem impacts our personal lives, our environment, and our economy. Surely, this large-scale problem will not be solved by one company or one community. A national strategy needs to be put into place that integrates parking data from a variety of sources and provides parking solutions for everyone.

Google decided to step in to help solve a small portion of this problem by adding a parking feature to its Google Maps app. That feature provides users with an estimation of the probability of finding a parking spot in a specific location as "easy," "medium," or "limited."

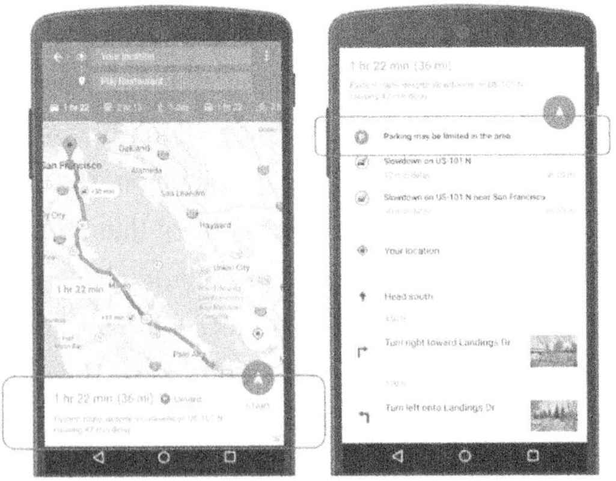

Although the feature appeared at first to be helpful, in reality it does not solve the problem, because Google Maps does not show which parking lots have available spaces and does not show which spaces are open in parking lots that do have available spaces. Furthermore, the app does not provide any information about parking meters on the street. Thus, it will not reduce the pollution that results from people driving around and around to find parking spaces.

The question here is, how does Google Maps know the probability of finding parking spaces in a certain city or location? The answer is that Google uses algorithms to calculate the time that you would take from the moment you arrive in your car in front of your destination until you have completely stopped moving. For instance, if I drive from my home in Long Beach to a location in Los Angeles, algorithms start counting how much time I spend driving around and walking after I arrive at the address I gave to Google Maps as my destination. The algorithms then record this time as time spent finding a parking spot.

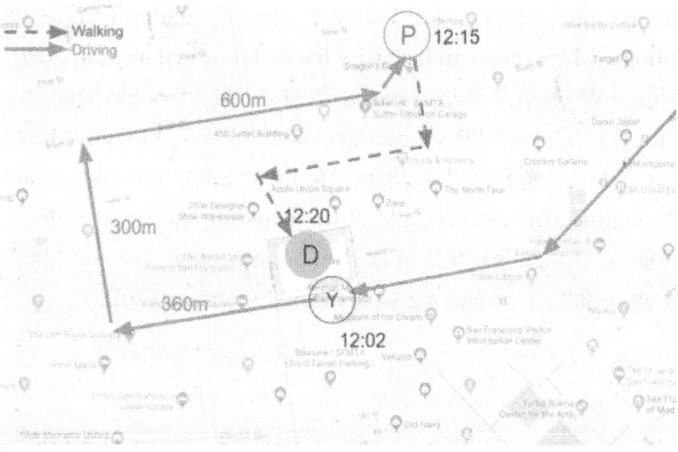

Source: Neha et al., 2019

Imagine you have a ghost following you wherever you go, which counts how many minutes you drive and how many steps you walk after you arrive in front of your destination until you are inside the building.

Google Maps then takes that data and aggregates it with past parking data from all other Google Maps users who have driven to that neighborhood, and then calculates average times and steps to park in that neighborhood on a particular day and time. The next time anyone asks Google Maps how long it will take to find a parking space in that neighborhood, Google Maps will use its aggregated data to predict parking availability there as "easy," "medium," or "limited."

But Google Maps does not stop there. Behind the scenes, it is also using that data to target people and businesses for ads.

The problem is that Google Maps' parking algorithms are not quite accurate. For instance, what happens to those of us who, after getting to our final destination, decide to drive around and explore the neighborhood before parking? We might have also decided to park a bit further away for a cheaper parking rate. There are any number of variations that algorithms cannot differentiate. Furthermore, this particular feature of Google Maps' app can be easily hacked. For instance, in 2020, Simon Weckert, a Berlin-based artist, tricked Google Maps by filling a cart with 99 smartphones and opening Google Maps' navigation app on *all* of them. He then slowly pulled the cart through Berlin's streets, playfully including the street right in front of Google's office in Berlin. Soon enough, Google Maps showed the street blazing with traffic (Shammas, 2020).

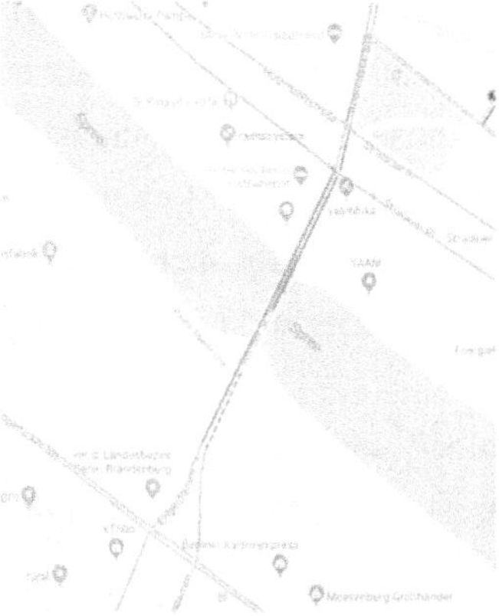

Source: https://www.washingtonpost .com

How many of us want to have someone tracking every single move we take daily? I am not denying that Google Maps provides a reliable navigation service, but what are we giving up in return? Who can regulate and protect our privacy? These are some questions that we should carefully consider.

Data Can Be Used for Good or for Evil

Obviously, data gathering can have socially beneficial purposes. For instance, data about people's usage of roads and bridges can help lawmakers to understand where they should spend taxpayers' money. In addition, data that analyzes energy usage can help utility companies to project how much energy they need to supply at different times of day, or in different seasons. However, there are ways that data gathering can be used for antisocial purposes.

In today's information age, many cities have begun the process of transitioning into becoming so-called "smart cities." To be accurate, this transition requires extensive data on the cities' residents. Thus, gathering such data means deploying tens of thousands of sensors and actuators across cities. In 2020, American cities collected 183,500 *types* of data about their citizens, and the number of those types was increasing by 2,791 every single day (Meiyi et al., 2020). Ominously, this exponential amassment of data continually increases the chances of bad actors identifying individuals and their behavior for malevolent purposes.

When this city data is combined with demographic data collected by the U.S. Census Bureau and other governmental agencies, and then that data is combined with the data collected by private businesses and social media, the potential damage to society becomes obvious.

But the problem is not with technology or the data collected. It is with the officials or executives who are responsible for managing and protecting the data. Current policies to protect citizens' privacy need to be enhanced and strengthened to match today's technological advances. The public needs to be made more aware of how their data is being used—and sometimes abused.

Cybersecurity Should Be Taught in Schools

When I recently asked my students, "How many devices connected to the Internet do you have at home?" I was shocked by their answers. The average was between forty and sixty devices! Those numbers included phones, TVs, computers, printers, Alexa, smart lights, and so on.

I managed to grow up without the benefit of cell phones, tablets, or the Internet. Today, however, especially during the pandemic, kids are being educated, are playing games, and are watching movies almost entirely online—unaware that their data privacy is being violated daily.

Kids need to know how to keep their devices protected. They need to be made aware of the basis behind engineered social media attacks, the main principles of data security and data privacy, and the significant differences between them.

In Israel, the government has made cybersecurity a national mission. In 2017, the Israeli government announced the establishment of a national center for cyber education, whose sole purpose is to teach students the basic building blocks of the web and cybersecurity. Earlier, I mentioned how and why the Internet is not secure and why that lack of security will be an ongoing problem in the future.

Now that we have entered the Fourth Industrial Revolution, there will be far more automated services that will make our lives easier. We are now seeing the likes of autonomous cars, smart homes, and even robots that can perform surgery. However, along with these technological marvels, harmful security threats are constantly spawning like parasites. For that reason alone, we must be well informed about our data security and data privacy, so that we know how our data is being used and misused. It is not only incumbent on us to be aware of these matters, but also to prepare the next generations to be effective leaders in a future where the Internet will be essential to everyone for their survival.

Chapter 10
What Does It All Mean?

One day while browsing my Facebook feeds, I was prompted to watch a video by Steve Harvey, who began with:

> Can I ask you a question? Do all of you have a Smartphone? Has your Smartphone done to you what it's done to me? Has your Smartphone made you stupid? I am actually stupid now because of my Smartphone. I know nothing now. I used to know people's phone numbers. I used to know my mom's number and my friend's number, but now I don't even know my wife's number!

So, has technology made us smarter or stupider? Believe it or not, that question has been asked for centuries. A Swiss scientist once raised the concern that modernity had engulfed people with data that had become "confusing and harmful" to the mind. His concern was directed at the overwhelming data emerging from the printing press. The scientist, Conrad Gessner, died in 1565 (Mellen, 2016).

The same question about technology was brought up as far back as 399 B.C., when Socrates famously cautioned against writing because of the adverse effects it supposedly has on the minds of children. Writing, Socrates asserted, "creates forgetfulness in the learners' souls because henceforth they would not use their memories" (Berry, n.d.). In my observation, technology is not making us stupid. It is simply redefining the way we think and how we utilize our memory.

Consider the difference between the human brain and a computer algorithm. The algorithm runs faster and produces far more results than a human brain can. Of course, the algorithm doesn't have feelings, nor can it sense or intuit things that are intrinsically human, such as hunger and sex drive.

By increasing the resources that an algorithm needs, such as Random Access Memory (RAM) or the Central Processing Unit (CPU), one can enable an algorithm to greatly outpace the human brain, since our gray matter only operates within a limited capacity. We cannot make our minds function faster by adding RAM, CPU, or whatever. We can, however, maximize our mental output by eliminating superfluous elements to help us focus on performing productive tasks. For instance, I can free some of my brain memory by ignoring the need to memorize phone numbers. Instead, I can utilize my memory by storing information about an upcoming exam. I can also memorize a new skill set that I just learned. To conclude that human beings have become less intelligent because we are utilizing current technology is shortsighted.

Technology such as the Smartphone is nothing more than another version of human innovation, which dates back thousands of years. Innovations appeared long before the First

Industrial Revolution. They include the writing system of the ancient Sumerians, the mummification process invented by the Egyptians, the numbering system of the Babylonians, and the monetary system of the ancient Greeks.

Human beings inherently cling to whatever is the tried and true. To use the present-day vernacular, we like to stay in our "comfort zone," so our natural curiosity has always been countered with caution. Hesitation to make changes has often led to tension. To make matters worse, for centuries our suspicious nature has given rise to many superstitions. Such false beliefs made it slow going for a long time. Witness the Dark Ages—some 900 years of intellectual stagnation.

The changes brought about by Industry 4.0 have only just begun. As time passes, old jobs will disappear and new ones will be created even faster. Individuals, companies, cities, and whole nations will have to overhaul their way of doing things to conform to Industry 4.0. All our lives have been and will continue to be impacted, whether we like it or not.

The focus in the future will be far more on automation, customization, high productivity, and efficiency. Of course, this was also true of the previous industrial revolutions, but the difference today is that, to borrow a phrase from Star Trek, things will be changing at "warp speed."

Coping with Industry 4.0

Long ago, our ancestors discovered agriculture as a source of food instead of relying on hunting and gathering. With that discovery, we went from being a migratory species to one based on stationary farming. Things pretty much remained that way for about 12,000 years. Then we underwent the First Industrial Revolution. In the past 300 years, we have undergone different versions of industrial revolutions.

Has technology changed our lives for the better? Of course, it has. However, it also came with challenges that were not solely the result of technological advances. Rather, those challenges have always stemmed from how we have reacted to and resisted each successive level of technology that has come along.

In this book, I have pointed out examples of Industry 4.0 technologies that have made it possible to resolve any number of the problems we struggle with every day. On the other hand, I have also noted that many of our policymakers do not have even a passing knowledge of today's technologies, and that ignorance has affected us deeply.

As citizens, we have a responsibility to stay informed about technology's benefits and how they are impacting our daily lives. I have already underscored that point several times. However, no one technology or system will serve as a panacea for all our problems, because everything now is and will forever be changing—and faster and faster over time. To keep up with what is taking place today, we need to be technologically agnostic and emotionally intelligent.

Technological Agnosticism

To be technologically agnostic is to constantly have an eye out for new tools and technologies that will work better than our current ones. That awareness is not limited to software or hardware, but can apply to anything in life, from personal lifestyles to business procedures. Technology has never been more than a tool to make our lives easier, so we need to see that this tool constantly improves with time. It is important that we always stay open to learning, applying, and leveraging new technologies, which we can use in so many meaningful ways to help us achieve our desired results.

This same principle applies to our jobs. If we recognize that our job can be done more efficiently by using some form of technology, we should at least advocate for that change and, if possible, lead it. If that is impractical, it is time to move on to the next job.

It is obvious these days that if we don't take our job security into our own hands, our employers will inevitably let us go. Keep in mind that business decisions are not directed at you personally. It's just how the system is set up. Companies will always look for cheaper ways to increase profits, regardless of the human cost.

In Chapter 8, I explained how our mindset plays a huge role in our adapting to or resisting change. But changing our mindset is easier said than done, since it requires continuous learning and growing. Above all, flexibility is the key to adapting to nature's primary constant: change. The good news is that we all have greater or lesser degrees of a growth mindset. Suffice it to say that it has never been more important to use the growth mindset than now!

Emotional Intelligence

Emotional intelligence refers to our ability to perceive, control, evaluate, and express emotions. Although automation is one of the main thrusts of Industry 4.0, our emotional and social intelligence cannot be automated as of yet. Many experts believe that the emotional intelligence quotient (EQ) may be more important than our intelligence quotient (IQ). An individual's EQ will be one of the main skill sets required for most upcoming jobs. EQs include self-awareness, self-regulation, self-motivation, empathy, and social skills.

Self-awareness is a prime virtue that includes not only being able to understand our own strengths and weaknesses, but also being able to learn and grow from constructive criticism.

Self-regulation includes the ability to maturely express our emotions with restraint and control.

Self-motivation is the drive to pursue our goals by completing tasks that are inspired by our inner ability to take action. It is that force within us that is not motivated only by the acquisition of money or prestige.

Empathy is the ability to have compassion and understanding for others, so that we connect with them on an emotional level.

Social skills are the set of abilities that enable us to build relationships with others based on trust and respect, while avoiding power struggles and backstabbing.

Putting Things into Perspective

My intention in this book has been to help nurture the right culture in our society by awakening my readers to the potential impacts of the technological advances that we are currently living with. Given my academic and technical background, I can "see" what lies just ahead. Aware, however, that most people have no idea what is just around the corner, I feel impelled to inform my readers about the dislocations as well as the benefits brought about by the digital transformations of Industry 4.0.

Like characters in a science fiction novel, we are unwittingly being engineered into an intelligence-based social hierarchy. Those of us who keep learning and growing technologically will reap the benefits to come. Those who resist inevitable change will fall on hard times.

Without deepening our understanding of today's technology, we will remain unaware that such things as our daily problems with parking, traffic, and pollution can be resolved. Without broadening our perspectives, we cannot entirely hold our public officials accountable for inept leadership, much less expect them to improve our standard of living. If we continue to live in technological ignorance, we will become victims of an automation that is fast overtaking all manual labor. If we remain technologically unenlightened, we will be at the mercy of companies that will quickly replace our services with robots or artificial intelligence.

Pessimistic prophesies of this nature are to be found in books written by and for professional persons who hold graduate degrees. In this book, however, I have deliberately presented the information as concisely and as straightforwardly as I could so as to reach the widest possible audience.

I am sounding this clarion call because as many people as possible need to be brought up to speed about just how much present-day technology is reshaping our day-to-day lives. I am reaching out to high school students, the unemployed, factory workers, service workers, specialists, professionals, and anyone else who is living and breathing on this planet in the twenty-first century.

We are all in this together.

References

Agarwal, V., Eloot, K., & Patel, A. (2020, December 16). *Moving Past the "Pilot Trap" to Unleash Industry 4.0 in Indonesia*. Available at https://www.mckinsey.com/business-functions/operations/our-insights/moving-past-the-pilot-trap-to-unleash-industry-4-0-in-indonesia/.

Aristotle. (2021, February 2). *Categories*. Available at https://plato.stanford.edu/entries/aristotle-categories/.

Arora, N., Cook, J., Kumar, R., Kuznetsov, I., Li, Y., Liang, H. J., Miller, A., Tomkins, A., Tsogsuren, I., & Wang, Y. (2019). *Hard to Park?* Proceedings of the 25th ACM SIGKDD International Conference on Knowledge Discovery & Data Mining. Available at https://doi.org/10.1145/3292500.3330767

Ayaz, M., Ammad-Uddin, M., Sharif, Z., Mansour, A., & Aggoune, E. H. M. (2019). *Internet-of-Things (IoT)-Based Smart Agriculture: Toward Making the Fields Talk*. Available at https://doi.org/10.1109/access.2019.2932609

Berry, C. (n.d.). *Technology of the Book Web Ring*. Available at https://cla.purdue.edu/academic/english/theory/webring/berryPlato.html/.

Bhisey, R. (2020, October 26). *Synthetic Biology Market Is Projected to Expand at a CAGR of 26.3% from 2019 to 2027. BioSpace*. Available at https://www.biospace.com/article/synthetic-biology-market-is-projected-to-expand-at-a-cagr-of-26-3-percent-from-2019-to-2027/.

Bienhaus, D. (2017). *Patterns for the Industrial Internet/Industry 4.0*. Proceedings of the 22nd European Conference on Pattern Languages of Programs. doi:10.1145/3147704.3147723/.

Boudette, E. (2020, August 4). *Jim Hackett, Ford's New C.E.O., Has a History of Turnaround Stories*. Available at https://www.nytimes.com/2017/05/22/business/jim-hackett-ford.html/.

Burrows, L. (2013, July 24). *To Be Let Alone: Brandeis Foresaw Privacy Problems*. Available at https://www.brandeis.edu/now/2013/july/privacy.html/.

Cambridge University. (2016, November 25). *The Evolution of a Misquotation. Darwin Correspondence Project*. Available at https://www.darwinproject.ac.uk/people/about-darwin/six-things-darwin-never-said/evolution-misquotation/.

Carmichael, J., & Carmichael, F. (2020, May 30). *Coronavirus: Bill Gates "Microchip? Conspiracy Theory and Other Vaccine Claims Fact-Checked*. Available at https://www.bbc.com/news/52847648/.

City of Long Beach. (n.d.). *Long Beach Smart City*. Available at https://longbeach.gov/smartcity/guiding-principles/strategy/.

City of San Diego. (n.d.). *Smart City, Sustainability*. Available at https://www.sandiego.gov/sustainability/smart-city/.

Colom, R., Karama, S., Jung, R., & Haier, R. (2010, December 1). *Human Intelligence and Brain Networks*. Available at https://www.ncbi.nlm.nih.gov/pmc/articles/PMC3181994/.

Curry, A. (2021, May 26). *Obama on the Economy.* Available at https://www.today.com/video/obama-on-the-economy-and-weiner-44518979712/.

Dignan, L. (2017, October 3). *Ford CEO Hackett: Digital Transformation Means Making the Business Fit First.* Available at https://www.zdnet.com/article/ford-ceo-hackett-digital-transformation-means-making-the-business-fit-first/.

Dweck, C. S. (2007). *Mindset: The New Psychology of Success*. New York: Ballantine Books.

European Parliament. (2014). *Mapping Smart Cities in the EU*. Available at https://www.europarl.europa.eu/RegData/etudes/etudes/join/2014/507480/IPOL-ITRE_ET%282014%29507480_EN.pdf/.

Elmjid, F. (2017, October 23). *World Population Projected to Reach 9.8 Billion in 2050, and 11.2 Billion in 2100.* Available at https://www.un.org/development/desa/en/news/population/world-population-prospects-2017.html/.

Geier, D. (2019, November 22). *The Best Leaders Lead By Example.* Available at https://medium.com/swlh/the-best-leaders-lead-by-example-88ee68186aae#:%7E:text=Show%20others%20that%20how%20you,something%3B%20they%20do%20it.%E2%80%9D&text=John%20Wooden%2C%20the%20head%20basketball,with%20people%20of%20good%20character/.

Green, C. (2021, April 23). *History of the PLC.* Available at https://library.automationdirect.com/history-of-the-plc/.

Guilford-Blake, R. (2020, February 18). *Wait. Will AI Replace Radiologists After All?* Available at https://www.radiologybusiness.com/topics/artificial-intelligence/wait-will-ai-replace-radiologists-after-all/.

Hall, R. E., Bowerman, B., Braverman, J., Taylor, J., Todosow, H., & Wimmersperg, V. U. (2000, September 28). *The Vision of a Smart City*. Available at https://www.osti.gov/biblio/773961/.

Harari, Y. N. (2017). *Homo Deus: A Brief History of Tomorrow*. New York: HarperCollins.

Harrison, C., Eckman, B., Hamilton, R., Hartswick, P., Kalagnanam, J., Paraszczak, J., & Williams, P. (2010). "Foundations for Smarter Cities." *IBM Journal of Research and Development*, *54*(4):1–16.

Herald Globe. (2014, July 12). *City Population to Reach 6.4 Billion by 2050*. Available at https://www.heraldglobe.com/news/223727231/city-population-to-reach-64bn-by-2050/.

Hidayatno, A., Rahman, I., & Daniyasti, D. (2019). *Conceptualizing the Promise of Industry 4.0*. Available at https://doi.org/10.1145/3364335.3364350/.

Hirsch, B. T. (2008, November 1). *Sluggish Institutions in a Dynamic World: Can Unions and Industrial Competition Coexist?* Available at https://www.aeaweb.org/articles?id=10.1257/jep.22.1.153/.

History, C. (2020). *The Industrial Revolution: A Captivating Guide to a Period of Major Industrialization and the Introduction of the Spinning Jenny, the Cotton Gin, Electricity, and Other Inventions*. Chicago: Captivating History.

Inrix Analytics Company. (2017, July 12). *Searching for Parking Costs*. Available at https://inrix.com/press-releases/parking-pain-us/.

International Business Machines. (2010). *Smarter Cities for Smarter Growth: How Cities Can Optimize Their Systems for the Talent-Based Economy*. New York: IBM Global Business Services.

Kane, G. C., Phillips, A. N., Copulsky, J. R., & Andrus, G. R. (2019). *The Technology Fallacy: How People Are the Real Key to Digital Transformation (Management on the Cutting Edge)*. Cambridge, MA: MIT Press.

Keiser, K. *The Real Cost of Parking Tickets: 16 U.S. Cities Make Over $1.4 Billion in Annual Parking Ticket Revenue*. Available at www.carrentals.com/blog/parking-tickets-cost-americans/.

Kremen, G. (1974). MDTA: *The Origins of the Manpower Development and Training Act of 1962*. Available at www.dol.gov/general/aboutdol/history/mono-mdtatext/.

Liao, Y., Deschamps, F., Loures, E. D. F. R., & Ramos, L. F. P. (2017). *Past, Present, and Future of Industry 4.0: A Systematic Literature Review and Research Agenda Proposal*. Available at https://doi.org/10.1080/00207543.2017.1308576/.

McKinsey. (2020, January 24). *Unlocking Success in Digital Transformations*. Available at https://www.mckinsey.com/business-functions/organization/our-insights/unlocking-success-in-digital-transformations#/.

Meiyi Ma, S. M., Preum, M. Y., Ahmed, W. T., Abdeltawab, H., & Stankovic, J. A. (2020, February). *Data Sets, Modeling, and Decision-Making in Smart Cities: A Survey*. Available at DOI:https://doi.org/10.1145/3355283/.

Mellen, M. (2016, December 15). *People Have Always Been Afraid of the Effects of New Technology*. Available at https://www.mickmel.com/people-always-afraid-effects-new-technology/#:%7E:text=There%20was%20Conrad%20Gessner%2C%20who, due%20to%20the%20printing%20press/.

Monteleone, S., de Moraes, E. A., & Maia, R. F. (2019). *Analysis of the Variables That Affect the Intention to Adopt Precision Agriculture for Smart Water Management in Agriculture 4.0 Context*. Available at https://doi.org/10.1109/giots.2019.8766384/.

Morrison, S. (2021, February 1). *Facebook and Apple Are Fighting Over Privacy: What That Means to You*. Available at https://www.vox.com/recode/22254815/facebook-apple-privacy-ios-14-lawsuit/.

Mossberger, K., Clarke, S. E., & John, P. (2015). *The Oxford Handbook of Urban Politics*. New York: Oxford University Press.

Murray, S. (2020, October 29). *IDC Reveals 2021 Worldwide Digital Transformation Predictions; 65% of Global GDP Digitalized by 2022*. Available at https://www.idc.com/getdoc.jsp?containerId=prUS46967420/.

Nath, H., Apte, U., & Karmarkar, U. (2020). *Service Industrialization, Employment and Wages in the U.S. Information Economy*. Available at https://doi.org/10.1561/0200000050/.

National Geographic Society. (n.d.). *Impact of Technology on Agriculture.* Available at https://www.nationalgeographic.org/topics/impact-technology-agriculture/?q=&page=1&per_page=25/.

National Institute of Food and Agriculture. (n.d.). *Farmer Education.* Available at https://nifa.usda.gov/topic/farmer-education/.

Norman, J. (n.d.). *In the Swing Riots Agricultural Workers Destroy Threshing Machines.* Available at www.historyofinformation.com/detail.php?id=4655/.

Open Group, The. (n.d.). *The TOGAF Standard, Version 9.2: Architecture Principles.* Available at https://pubs.opengroup.org/architecture/togaf9-doc/arch/chap20.html#:%7E:text=Principles%20are%20general%20rules%20and,sets%20about%20fulfilling%20its%20mission/.

Peters, G., & Woolley, J. (1928, June 12). *Republican Party Platform of 1928: The American Presidency Project.* Available at https://www.presidency.ucsb.edu/documents/republican-party-platform-1928/.

Rajamanickam, V. (2019, March 26). *JD.com Opens Automated Warehouse That Employs Four People But Fulfills 200,000 Packages Daily.* Available at https://www.freightwaves.com/news/technology/jdcom-opens-automated-warehouse-that-employs-four-people-but-fulfills-200000-packages-daily/.

Reich, R. B. (2015). *Saving Capitalism: For the Many, Not the Few*. New York: Knopf.

Robnett, J. C. (2020, November 24). *J. C. R. Licklider and the Universal Network*. Available at https://www.livinginternet.com/i/ii_licklider.htm/.

Rogers, C. (2021, January 29). *Apple's Tim Cook: Business Cannot Turn a Blind Eye to Data Exploitation*. Available at https://www.marketingweek.com/apple-data-privacy/.

Rosetta. (2021, April 30). *How Long Does It Take To Get Test Results for A Procedure?* Available at https://rosettaradiology.com/rosetta-radiology-blog/how-long-does-it-take-to-get-test-results-for-a-procedure/.

Schwab, K. (2017). *The Fourth Industrial Revolution*. Redfern, Australia: Currency.

Segall, L. (2018, March 22). *Zuckerberg: I'm Really Sorry That This Happened*. Available at https://edition.cnn.com/videos/cnnmoney/2018/03/22/zuckerberg-facebook-sorry-cambridge-analytica.cnn/.

Serling, R., & Donner, R. (1964, May 15). "The Brain Center at Whipple's." Season 1, Episode 2, *The Twilight Zone*. Los Angeles: Cayuga Productions, Inc.

Shammas, B. (2020, February 5). *A Man Walked Down a Street with 99 Phones in a Wagon. Google Maps Thought It Was a Traffic Jam*. Available at www.washingtonpost.com/technology/2020/02/04/google-maps-simon-weckert/.

Slynchuk, A. (2021, May 5). *Big Brother Brands Report: Which Companies Track Our Personal Data the Most?* Available at https://clario.co/blog/which-company-uses-most-data/.

Smart Cities Council. (n.d.). *The Definition of a Smart City*. Available at https://rg.smartcitiescouncil.com/readiness-guide/article/definition-definition-smart-city/.

U.S. Air Force. (2019, July 9). *Donovan Stresses "Digital Air Force's" Importance and Necessity*. Available at https://www.af.mil/News/Article-Display/Article/1899838/donovan-stresses-digital-air-forces-importance-and-necessity/.

Voss, C., & Raz, T. (2016). *Never Split the Difference: Negotiating As If Your Life Depended On It*. New York: Harper Business.

Weareclario.com. (2021, May 5). *Big Brother Brands Report: Which Companies Track Our Personal Data the Most?* Available at https://clario.co/blog/which-company-uses-most-data/.

Zachman, J. A. (2008). *The Concise Definition of the Zachman Framework*. Available at https://www.zachman.com/about-the-zachman-framework/.

Zone-H. (2021, June 23). *Zone-H's Cybercrime Archive*. Available at http://www.zone-h.org/archive/special=1/.

Made in the USA
Monee, IL
24 March 2026

46751543R00105